Carbon Nanotube and Graphene Nanoribbon Interconnects

Carbon Nanotube and Graphene Nanoribbon Interconnects

Debaprasad Das

Hafizur Rahaman

CRC Press
Taylor & Francis Group
Boca Raton London New York

CRC Press is an imprint of the
Taylor & Francis Group, an **informa** business

CRC Press
Taylor & Francis Group
6000 Broken Sound Parkway NW, Suite 300
Boca Raton, FL 33487-2742

First issued in paperback 2017

ISBN-13: 978-1-4822-3948-5 (hbk)
ISBN-13: 978-1-138-82231-3 (pbk)

Library of Congress Cataloging-in-Publication Data

Das, Debaprasad.
    Carbon nanotube and graphene nanoribbon interconnects / authors, Debaprasad Das, Hafizur Rahaman.
        pages cm
    Includes bibliographical references and index.
    ISBN 978-1-4822-3948-5 (hardcover : alk. paper) 1. Interconnects (Integrated circuit technology)--Materials. 2. Nanotubes. 3. Nanoelectronics--Materials. I. Rahaman, Hafizur. II. Title.

    TK7874.53.D37 2015
    621.39'5--dc23   2015004560

Visit the Taylor & Francis Web site at
http://www.taylorandfrancis.com

and the CRC Press Web site at
http://www.crcpress.com

To my parents, my wife, Joyita, and daughters, Adrija and Adrisha

Debaprasad Das

To my parents, my wife, Anju, and son, Rohan

Hafizur Rahaman

Contents

Preface

The aggressive scaling of very large-scale integration (VLSI) technology has pushed the device and interconnect dimensions of the integrated circuit (IC) to the nanometric dimensions. This has resulted in many low-dimensional issues that show serious threats to the further advancement of the VLSI technology. As predicted by the International Technology Roadmap for Semiconductors (ITRS) and also supported by many research articles, carbon nanotube (CNT) and graphene nanoribbon (GNR) could become potential replacements for the traditional CMOS devices and the copper-based interconnect systems. This book provides a comprehensive analysis of the CNT- and GNR-based VLSI interconnects at the nanometric dimensions to prove their benefits. The comprehensive analytical model for interconnects and use of CNT and GNR are beneficial to VLSI designers working in this area. Starting with a brief introduction about the carbon nanomaterials, this book first discusses the research works carried out in this area by different research groups. Then, it presents details about the model of the CNT and GNR interconnects. The applicability of CNT and GNR is studied by modeling them as signal and power interconnects and results are presented in detail in this book. RF analysis and stability of the CNT and GNR interconnects are also presented. The book concludes with the potential applications of the CNT and GNR.

Debaprasad Das
Assam University

Hafizur Rahaman
Indian Institute of Engineering Science and Technology

MATLAB® is a registered trademark of The MathWorks, Inc. For product information, please contact:

The MathWorks, Inc.
3 Apple Hill Drive
Natick, MA 01760-2098 USA
Tel: 508-647-7000
Fax: 508-647-7001
E-mail: info@mathworks.com
Web: www.mathworks.com

Acknowledgments

We are grateful to Dr. Gagandeep Singh of CRC Press for encouraging us to pursue this book project. We convey our sincere gratitude to our respected professor Bhargab B. Bhattacharya for his kind support, inspiration, and guidance throughout our research career. My sincere thanks go to all the peer reviewers of the book.

Our sincere thanks go to our colleagues and friends for all their support that was needed for our work. We are grateful to our wives and our children and other family members for their constant support and inspiration.

We sincerely apologize to the readers for any unintentional mistakes that may have crept into this book. All suggestions and feedback for further improvement of the book are welcome.

Debaprasad Das
Assam University

Hafizur Rahaman
Indian Institute of Engineering Science and Technology

Authors

Dr. Debaprasad Das was born in Haria, Purba Medinipur, West Bengal, India, on May 10, 1975. He received a bachelor's (honors) degree in physics in 1995, a bachelor's degree in radio physics and electronics in 1998, a master's degree in electronics and telecommunication engineering in 2006, and a PhD in engineering from the Vidyasagar University, University of Calcutta, Jadavpur University, and Bengal Engineering and Science University, Shibpur, respectively. He was with the ASIC Product Development Centre, Texas Instruments, Bangalore, as a senior engineer from 1998 to 2003. He worked in the Department of Electronics and Communication Engineering, Meghnad Saha Institute of Technology, Kolkata, India, from 2003 as an assistant professor and is a member of IEEE. Presently, he is working as an associate professor and head in the Department of Electronics and Telecommunication Engineering, Assam University, Silchar, India. He has authored or coauthored several research papers in national and international journals and conferences.

Dr. Das is the author of the book *VLSI Design*, published in 2010. He also authored *Timing and Signal Integrity Issues with VLSI Interconnects and Digital Logic Design using Carbon Nanotube Field Effect*, published by Lambert Academic Publishing, Germany. His research interests include VLSI design, developing of EDA tools for interconnect modeling, analyzing crosstalk and reliability, digital CMOS logic design, and modeling of nanoelectronic devices and interconnects. Dr. Das received a gold medal from Vidyasagar University in 1997 for emerging first in the undergraduate degree. He received a First Place Award in the PhD forum at VDAT 2012.

Dr. Hafizur Rahaman received a bachelor's degree in electrical engineering from Bengal Engineering College, India, in 1986, and a master's degree in electrical engineering and a PhD in computer science and engineering from Jadavpur University, Kolkata, India, in 1988 and 2003, respectively. He is a full professor of the Indian Institute of Engineering Science and Technology, Shibpur, West Bengal, India. He received an INSA-Royal Society UK Fellowship Award during 2006–2007. He also received the Royal Society (UK) International Fellowship Award during 2008–2009 to conduct advanced research in the area of VLSI and nanotechnology at University of Bristol, United Kingdom.

His research interests include VLSI design and testing, CAD for microfluidic biochips, emerging nanotechnologies, and reversible computing. He has published more than 280 research articles in archival journals and refereed conference proceedings. Recently, he received a DST-DAAD research fellowship award under the Indo-German (Department of Science Technology,

India-DAAD) Bilateral Cooperation in 2013. He leads the VLSI design and test group at his institute. He is a member of the VLSI Society of India (VSI) and ACM SIGDA and a senior member of IEEE. Dr. Rahaman has served as PC chair at VDAT 2012, VDAT 2014 and ISED 2012; served in program committees of numerous conferences such as VLSI Design, ASP-DAC, ATS, and ISVLSI; and organized several seminars, tutorials, workshops, and several technical sessions.

1

Introduction to Allotropes of Carbon Nanomaterials

1.1 Introduction to Carbon Nanotube and Graphene Nanoribbon

Carbon nanotube (CNT) and graphene nanoribbon (GNR) are two new important carbon nanomaterials in the nanometer regime. Though CNT was discovered more than two decades ago, in 1991, by Japanese physicist Sumio Iijima, GNR was discovered more recently, in 2004, by Andre Geim and Konstantin Novoselov at the University of Manchester. Since their discovery, both of these carbon nanomaterials have gained a lot of importance due to their remarkable properties. Significant progress has been made in finding out the fundamental properties, exploring the possibility of engineering applications, and growth technologies. This chapter provides a brief description about CNT and GNR.

1.2 Graphene

A single layer of three-dimensional (3D) graphite forms a two-dimensional (2D) material called 2D graphite or a graphene layer. Graphene is an allotrope of carbon. Its structure is one-atom-thick planar sheets of sp^2-bonded carbon atoms that are densely packed in a honeycomb crystal lattice [1]. The term *graphene* was coined as a combination of *graphite* and the suffix *-ene* [2]. Graphene is most easily visualized as an atomic-scale chicken net made of carbon atoms and their bonds (Figure 1.1).

The carbon–carbon bond length in graphene is 0.142 nm [3]. Graphene sheets stack to form graphite with an interplanar spacing of 0.335 nm [4]. Graphene is the basic structural element of some carbon allotropes including graphite, charcoal, CNTs, and fullerenes. The band theory of 2D graphite or a graphene layer was studied more than six decades ago, in 1947 by Wallace [5]. But until 2004,

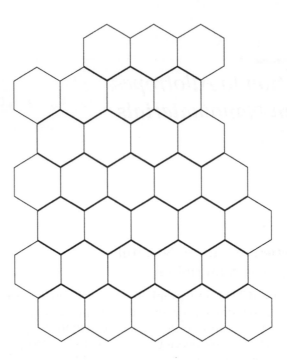

FIGURE 1.1
Graphene sheet.

it was believed that 2D crystals were thermodynamically unstable and could not exist. However, the experimental discovery of graphene in 2004 flaunted common wisdom, and the Nobel Prize in physics for 2010 was awarded to Andre Geim and Konstantin Novoselov at the University of Manchester "for their groundbreaking experiments regarding the two-dimensional material graphene" [6].

1.3 Graphene Nanoribbon

GNRs (also called nanographene ribbons) are strips of graphene with ultra-thin width (<50 nm). The theoretical study on electronic states of graphene ribbons was introduced as a theoretical model by Fujita et al. [7].

Depending on the orientation of carbon atoms on the edge of the graphene sheet, GNR is either armchair or zigzag (Figure 1.2). Zigzag GNR is always metallic, whereas armchair GNR can be either semiconducting or metallic depending on geometry (chirality). For interconnect applications, zigzag GNR is proposed due to its metallic property.

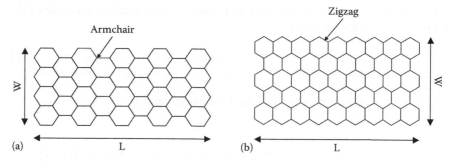

FIGURE 1.2
GNR with chirality (a) armchair and (b) zigzag.

1.4 Carbon Nanotube

CNT was discovered accidentally by Sumio Iijima in 1991. CNTs are basically rolled graphene sheets and show both semiconducting and metallic properties depending on their chirality. Although the semiconducting CNTs are being considered for nanoelectronic devices, the metallic CNTs are being considered for nanointerconnects. CNT can be either single-walled or multiwalled. The single-walled CNT (SWCNT) is a rolled graphene sheet, whereas the multiwalled CNT (MWCNT) is concentrically rolled graphene sheets. The diameter of SWCNTs varies from 0.7 to 5 nm, whereas that of the MWCNTs ranges from a few nanometers to tens of nanometers. Figure 1.3 shows a schematic of a graphene sheet.

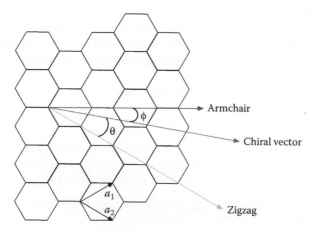

FIGURE 1.3
(See color insert.) Schematic of a graphene sheet.

The characteristic of a CNT is determined by its chirality. Chirality [3] is defined by a vector given by

$$\vec{P} = n\vec{a_1} + m\vec{a_2} \tag{1.1}$$

where:
$\vec{a_1}$ and $\vec{a_2}$ are the unit vectors of the hexagonal lattice
n, m are integers

The diameter of a CNT is determined by the pair (n, m) and is given by

$$d = \frac{\gamma}{\pi}\sqrt{n^2 + nm + m^2} \tag{1.2}$$

where γ is the length of unit vectors. The relation between γ and a_{CC} (carbon–carbon bond length) is given by

$$\gamma = \sqrt{3}a_{CC} \tag{1.3}$$

The value of a_{CC} is 0.142 nm.
    The chiral angle is defined as the angle between the vectors $\vec{P}$ and $\vec{a_1}$ [3].

$$\psi = \cos^{-1}\left[\frac{(2n+m)}{2\sqrt{n^2 + nm + m^2}}\right] \tag{1.4}$$

The chiral angle is used to separate CNTs into three classes differentiated by their electronic properties: armchair ($n = m$, $\psi = 30°$), zigzag ($m = 0$, $n > 0$, $\psi = 0°$), and chiral ($0 < |m| < n$, $0 < |\psi| < 30°$).
    Armchair CNTs are metallic (a degenerate semimetal with zero band gap). Zigzag and chiral nanotubes can be semimetals with a finite band gap if $(n-m) = 3k$ (k being an integer and $m \neq n$) or semiconductors in all other cases.
    The band gap for the semimetallic and semiconductor nanotubes scales approximately with the inverse of the tube diameter, giving each nanotube a unique electronic behavior.
    The dispersion relation in CNT is given by

$$\varepsilon(k) = v_F \hbar k \tag{1.5}$$

where v_F is the Fermi velocity (= 8×10^5 m/s) of electron in CNT. Figure 1.4 shows the electron energy (ε) versus wave vector (k) diagram.
    The separation between the quantum states is given by

$$\delta\varepsilon = \frac{d\varepsilon}{dk}\delta k = \hbar v_F \frac{2\pi}{l} = \frac{hv_F}{l} \tag{1.6}$$

where l is length of the one-dimensional (1D) system, that is, an isolated CNT. Due to the finite quantum energy level spacing of electrons in the 1D system,

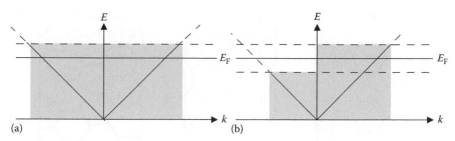

FIGURE 1.4
ε–k diagram for CNT.

there is an additional kinetic energy required to add an extra electron to the system. By equating this energy with an effective quantum capacitance e^2/C_Q, one can obtain the expression for quantum capacitance.

CNTs are basically classified into three types: (1) SWCNT, (2) double-walled CNT (DWCNT), and (3) MWCNT. SWCNT has only one graphene sheet rolled into cylinder. DWCNT and MWCNT have two and many concentric shells, respectively.

1.4.1 Single-Walled CNT

The SWCNT is basically a rolled graphene sheet in the form of a cylinder. The diameter of an SWCNT varies from 1 to 5 nm. It exhibits metallic or semiconducting properties depending on the chirality and can be classified into three types (based on the chirality): (1) armchair, (2) zigzag, and (3) chiral (Figure 1.5). Armchair SWCNTs are always metallic, whereas zigzag and chiral SWCNTs can be semiconducting or metallic. The metallic SWCNTs have long mean free path (MFP). Therefore, they are being considered for interconnect materials.

The semiconducting SWCNTs are being explored for making nanoelectronic devices. The band gap of semiconducting SWCNTs scales approximately with the inverse of the diameter of the tubes.

1.4.2 Multiwalled CNT

MWCNT is a CNT structure with several concentric shells of rolled graphene sheets. MWCNTs are classified into two types: (1) double-walled CNT (DWCNT) and (2) MWCNT. DWCNT and MWCNT have two and many concentric shells, respectively.

The diameter of MWCNT varies from several nanometers to tens of nanometers. Although SWCNTs can be either metallic or semiconducting depending on their chirality, MWCNTs are always metallic. Moreover, MWCNTs have similar current-carrying capacity (as metallic SWCNTs) but are easier to fabricate than SWCNTs due to easier control of the growth process.

Armchair   Zigzag   Chiral

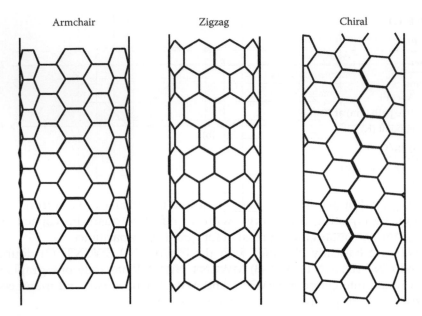

FIGURE 1.5
Different types of SWCNTs.

The circuit model of MWCNTs is much more complex than the traditional resistance, inductance, and capacitance (RLC) distributed interconnect model. Each shell in MWCNTs has different parameters and there are couplings between neighboring shells.

1.5 Properties of CNT

The important electrical, thermal, and mechanical properties of CNT and GNR are shown in Table 1.1.

Both CNT and GNR fundamentally originate from the 2D graphite. Graphite is an allotrope of carbon that is composed of 3D-layered hexagonal lattices of carbon atoms. A single layer of graphite is called 2D graphite or graphene. In graphite, each carbon atom has four valence electrons. Three of these electrons form strong bonds with neighboring atoms in the plane. These three covalent bonds are called σ bonds (2s, $2p_x$, and $2p_y$ orbitals). The fourth electron forms the $2p_z$ orbital perpendicular to the plane. The electrons forming the σ bonds do not contribute to the conductivity. The only electron that takes part in conduction is the fourth electron from the $2p_z$ orbital.

In graphene, each carbon atom has three σ bonds and one $2p_z$ orbital, which is perpendicular to the graphene plane, called the π covalent bond.

TABLE 1.1

Properties of CNT and GNR

Property	Tungsten	Copper	SWCNT	MWCNT	GNR
Maximum current density, J_{max} (A/cm²)	10^8	10^7	$>10^9$	$>10^9$	$>10^8$
Mean free path (nm) at 300 K	33	40	$>10^3$	2.5×10^4	1×10^3
Melting point (K)	3695	1357		3800	
Density (g/cm³)	19.25	8.94	1.3–1.4	1.75–2.1	2.09–2.33
Tensile strength (GPa)	1.51	0.22	22.2 ± 2.2	11–63	
Thermal conductivity ($\times 10^3$ W/m-K)	0.173	0.385	1.75–5.8	3	3–5
Temperature coefficient of resistance ($\times 10^{-3}$/K)	4.5	4	<1.1	−1.37	−1.47

The 2D energy dispersion relations for π bands of graphite, ε_{2D}, are given in [8]:

$$\varepsilon_{2D} = \pm \gamma_0 \left[1 + 4\cos\left(\frac{\sqrt{3}k_x a}{2}\right)\cos\left(\frac{k_y a}{2}\right) + 4\cos^2\left(\frac{k_y a}{2}\right) \right]^{1/2} \quad (1.7)$$

where γ_0 (= 3.033 eV) is the nearest-neighbor overlap integral.

The band structure of graphene is shown in Figure 1.6. The electrical properties of CNTs and GNRs are mainly due to the unique band structure of graphene. The ε–k (energy vs. wave vector) relation is linear for the low energies near the six corners of the 2D hexagonal Brillouin zone, which leads to zero effective mass for electrons and holes.

CNTs and GNRs have a very high current-carrying capability (at least 2 orders of magnitude higher than that of copper). CNTs and GNRs also

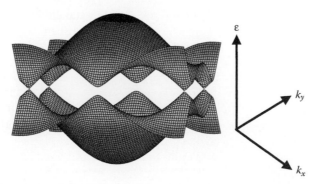

FIGURE 1.6
(See color insert.) ε–k relationship of GNR.

have long MFPs at low bias because of weak acoustic phonon scattering and suppressed optical phonon scattering at room temperature.

Both CNTs and GNRs can be formed from a single-layer graphene sheet. A GNR can be formed by cutting a ribbon out of graphene sheet following either armchair edge shape or zigzag edge shape. A CNT can be formed by rolling up a graphene sheet along circumferential vector given by $\vec{P}$ (see Equation 1.1).

The difference between CNT and GNR lies in their structural confinement conditions. The wave function along the circumference of CNT is periodic, whereas the wave function along the width of GNR vanishes at the two edges. These confinements are reflected in the band structure as slice cuts, each representing a subband. The position of the slice cuts depends on the chirality and diameter of CNT/GNR. Depending on the position of the slice cuts, the band structure could be without band gap (metallic) or with band gap (semiconducting).

It has been shown that CNT is metallic when $n - m = 3k$, where k is an integer [9]. Therefore, armchair CNTs are always metallic, whereas zigzag CNTs could be metallic or semiconducting, depending on the chiral indices (n, m). On the contrary, zigzag GNRs are always metallic and armchair GNRs can be either metallic or semiconducting, depending on the number (N) of atoms across the width: metallic when $N = 3k - 1$ and semiconducting when $N = 3k$ or $3k + 1$.

2

Growth of Carbon Nanotubes and Graphene Nanoribbon

2.1 Introduction

Carbon nanotube (CNT) and graphene nanoribbon (GNR) are two new materials made of carbon. Graphene is a two-dimensional structure where the carbon atoms are arranged in hexagonal array. The CNT is a rolled graphene sheet in the form of a cylinder. Both these materials have a great potential in making future very large-scale integration (VLSI) circuits due to their excellent properties. However, the main bottleneck of utilizing these materials in VLSI circuits is the suitable growth technology and compatibility with the existing silicon-based complimentary metal–oxide–semiconductor (CMOS) integrated circuit technology. These materials are synthesized using special techniques which are discussed in this chapter.

2.2 Works Related to CNT and GNR Technologies

The early growth processes of CNT were laser ablation and an arc-discharge approach. Both these processes can produce single-walled and multi-walled CNTs (SWCNTs, MWCNTs). The laser ablation technique can produce SWCNTs with very high purity (90%). However, it is not amenable for scale-up. The arc-discharge process can produce large quantities of SWCNTs but with modest purity. Chemical vapor deposition (CVD) has been widely used to grow CNTs in recent years. It has been shown that CVD is amenable for nanotube growth on patterned surfaces and for fabrication of electronic devices, sensors, field emitters, and other applications where controlled growth over masked areas is needed for further processing. More recently, plasma-enhanced CVD (PECVD) has been investigated for its ability to produce vertically aligned nanotubes.

The primary growth techniques of CNT are arc-discharge and CVD [1]. The arc-discharge method can produce high-quality CNT samples with diameter raging from 2 to 200 nm and length ranging from 1 to 100 µm [2]. But CNTs grown by this method suffer from two drawbacks when applied to interconnect applications. In this method, CNTs are produced in soot form and mixed with various other forms of carbon materials, which requires additional processing steps such as purification, sonication, and dispersion onto a substrate to be performed. In addition, nanotubes grown using this method are primarily laid horizontally on a flat substrate in a slow and random process. This restricts the nanotubes to side-contacted electron transport across a Schottky barrier formed at the metal–CNT junction. Therefore, a more robust and consistent architecture for nanotube growth must be used for their applications in VLSI.

CVD growth technique can yield CNTs with promising electrical characteristics for integration into on-chip interconnect systems. Delzeit et al. [3] demonstrated the growth of MWCNTs on various substrates by thermal CVD using multilayered metal catalysts.

Huang et al. [4] demonstrated the growth of highly oriented MWCNTs on polished polycrystalline and single-crystal nickel substrates by plasma-enhanced hot filament CVD at temperatures below 666°C. They obtained CNTs of diameter ranging from 10 to 500 nm and 0.1 to 50 µm in length depending on growth conditions. Uniform films of well-aligned CNTs were developed using microwave PECVD by Bower et al. [5]. They showed that nanotubes can be grown on contoured surfaces and aligned in a direction always perpendicular to the local substrate surface. They obtained multi-walled nanotubes of about 30 nm in diameter and 12 µm in length. The density of nanotubes was 4.4×10^9 nanotubes/cm^2.

Yao et al. [6] measured the intrinsic high-field transport properties of metallic SWCNTs. They found out that individual nanotubes can carry currents with a density exceeding 10^9 A/cm^2.

Delzeit et al. [7] generated high-density plasma from a methane–hydrogen mixture in an inductively coupled plasma reactor and grown MWCNTs on silicon substrates with multilayered Al/Fe catalysts. They obtained nanotubes that are vertically aligned, and the alignment is better than the orientation commonly seen in thermally grown samples.

Zhang et al. [8] demonstrated the growth of well-aligned CNTs of diameter ranging from 100 to 150 nm, on nickel-deposited silicon wafers by thermal CVD of ethylenediamine precursor. The CNTs are vertically aligned at high density over large areas on the silicon surface. They found that the size and the density of the nanotubes were dependent on the thickness of the nickel film. The length of the nanotube array was controlled by varying the CVD time.

Delzeit et al. [9] used ion beam sputtering for the sequential deposition of metal multilayers on various substrates to control the density of SWCNTs synthesized by CVD. The diameter range obtained is from 0.9 to 2.7 nm with most tubes at 1.3 nm.

Cui et al. [10] described a molecular memory device with semiconducting SWCNTs constituting a channel of 150 nm in length. They achieved data storage by sweeping gate voltages in the range of 3 V, associated with a storage stability of more than 12 days at room temperature.

Li et al. [11] reported a bottom-up approach to integrate MWCNTs into multilevel interconnects in silicon integrated circuit manufacturing. MWCNTs are grown vertically from patterned catalyst spots using PECVD.

Li et al. [12] studied the electric transport properties of an individual vertical MWCNT *in situ* at room temperature in a scanning electron microscope chamber. Their observations imply a multichannel quasiballistic conducting behavior occurring in the MWCNTs with large diameter. This can be attributed to the participation of multiple walls in electrical transport and the large diameter of the MWCNTs.

Chen et al. [13] reported a novel approach to grow highly oriented, freestanding, and structured CNTs between two substrates, using microwave plasma chemical vapor deposition. They measured the overall resistance of a CNT bundle and two CNT-terminal contacts, and it is found to be about 14.7 kΩ.

Gomez-Rojas et al. [14] presented an in-depth study of the radio frequency (RF) response of the SWCNT. They performed S-parameters analysis and obtained ac response for the frequency range 30 kHz–6 GHz.

Rice et al. [15] measured the electrical response of an individual MWCNT and its contacts up to 24 GHz.

Plombon et al. [16] measured the kinetic inductance of SWCNT bundle. They observed that the high-frequency impedance scales with the number of tubes.

Close and Wong [17,18] of Standford University demonstrated the assembly and electrical characterization of MWCNT-based interconnects. They studied four different materials (Al, Au, Pd, and Ti) for metal contact and found that Au and Pd are best contact metals. This work is also cited in [19].

Wu et al. [20] presented an in-depth electrical characterization of contact resistance in carbon nanostructure via interconnects.

Harutyunyan et al. [21] demonstrated the chiral-selective growth CNTs. Using the technique, they achieved metallic fraction of 0.91. By varying the noble gas ambient during thermal annealing of the catalyst, and in combination with oxidative and reductive species, they controlled the metallic fraction of the grown nanotubes.

Patil et al. [22] demonstrated wafer-scale growth of SWCNTs. They transferred aligned CNTs from quartz wafers to silicon wafers and fabricated CNT field-effect transistors (CNTFETs) using these transferred CNTs.

Franklin and Chen [23] showed that the nanotube transistors maintain their performance as their channel length is scaled from 3 μm to 15 nm. The short-channel effects of the traditional metal–oxide–semiconductor (MOS) devices were not observed for nanotube transistors. They demonstrated the performance of a nanotube transistor with 20 nm channel and contact lengths, 10 μA on-current, 1×10^5 on/off current ratio, and 20 μS peak transconductance.

Li et al. [24] presented a method to extract the contact resistance and bulk resistivity of vertically grown carbon nanofibers (CNFs).

Chen et al. [25] demonstrated the first monolithic integration of graphene interconnects with industry-standard CMOS technology, as well as the first experimental results that compare the performance of high-speed on-chip graphene and MWCNT interconnects.

Chai et al. [26] showed that electrical contact to the CNT can be substantially improved using a graphitic interfacial layer catalyzed by a nickel layer.

Ward et al. [27] demonstrated a method to reduce the CNT interconnect sheet resistance. They showed that interconnects made with SWCNTs have lower resistance than those with MWCNTs. They achieved sheet resistance of 50 Ω/sq using CNT functionalization and alignment technique. It is also projected that the sheet resistance can further be reduced to 10 Ω/sq using their techniques.

2.3 Works on Modeling and Analysis of CNT- and GNR-Based Interconnects

Burke [28,29] first proposed the electrical equivalent circuit model for CNT using Luttinger liquid theory. Salahuddin et al. [30] presented a model for 3D conductors to 1D CNTs to develop a transmission-line model. Pop et al. [31,32] presented a temperature-dependent model for CNT considering acoustic and optical phonon scatterings.

Srivastava and Banerjee [33] analyzed the applicability of CNT bundles as interconnects for VLSI circuits. They developed a model to extract the electrical equivalent model of the CNT bundle-based interconnects. They studied local, intermediate, and global interconnects and found that CNT bundle has better performance over copper interconnects at the intermediate and global lengths.

Raychowdhury and Roy [34] presented a circuit-compatible resistance, inductance, and capacitance (RLC) model for metallic SWCNT interconnect and studied its performance. Bias-dependent model of CNT resistance is presented in their work.

Naeemi and Meindl [35] presented models for MWCNT. A model is developed to provide the number of conducting channels in MWCNT based on the diameter of MWCNT and temperature.

Naeemi and Meindl [36] designed and evaluated the performance of SWCNT interconnects at the local, intermediate, and global lengths.

Naeemi and Meindl [37] modeled the temperature coefficient of resistance for both SWCNT and MWCNT interconnects.

Naeemi and Meindl [38] analyzed the performance of SWCNT- and MWCNT-based signal and power interconnects.

Massoud and Nieuwoudt [39] presented a comprehensive modeling of CNT-based interconnects.

Nieuwoudt and Massoud [40] modeled the magnetic inductance of the CNT bundle accurately using the method of partial element equivalent circuit (PEEC).

Nieuwoudt and Massoud [41] modeled the resistance of CNT bundle and developed the diameter-dependent models of contact and ohmic resistance. Nieuwoudt and Massoud [42] investigated the impact of process variations on future SWCNT bundle-based interconnects. They modeled 10 different sources of process variations. Nieuwoudt and Massoud [43] investigated the impact of magnetic and kinetic inductance on both delay and voltage overshoot. Their investigation revealed that kinetic inductance will not become significant in future SWCNT bundle-based interconnect systems.

Nieuwoudt and Massoud [44] presented comprehensive modeling and design techniques for CNT-based interconnects. Here, the authors modeled the electrostatic ground and coupling capacitances of MWCNT bundle accurately. They also found that large diameter MWCNT bundles are more susceptible to process variations than SWCNT bundles.

Haruehanroengra and Wang [45] analyzed the conductance of mixed SWCNT and MWCNT bundles. Their work demonstrates that the mixed CNT bundles can provide two to five times conductance improvement over copper by selecting the suitable parameters such as bundle width, tube density, and metallic tube ratio.

Rossi et al. [46] performed crosstalk analysis in CNT bus architecture. Their work showed that crosstalk-induced delay can be reduced up to 59% and crosstalk-induced voltage peak can be reduced up to 81% using their proposed CNT bus architecture.

Koo et al. [47] compared the performance of optical interconnect and CNTs. They showed that at lower bandwidth density and switching activity lower than 20%, CNT is most energy efficient provided its mean free path (MFP) and packing density are improved.

Li et al. [48] presented a compact model of MWCNT interconnects and investigated their performance. They also compared the performance of MWCNT-based interconnects with that of the traditional copper-based interconnects and SWCNT-based interconnects and found that MWCNT interconnects have smaller delay than that of copper interconnects at the intermediate and global lengths.

Srivastava et al. [49] studied the applicability of SWCNT-based interconnects in nanoscale integrated circuits. They analyzed the CNT-based vias for the first time. They showed that a densely packed CNT bundle can achieve 4× and 8× reduction in power compared to copper interconnects at 22-nm and 14-nm technology nodes, respectively.

Chen et al. [50] presented the results of electrothermal characterization of metallic SWCNT interconnect array. They investigated the self-heating impact on signal integrity of SWCNT interconnect array.

Pu et al. [51] performed crosstalk analysis in SWCNT and double-walled carbon nanotube (DWCNT) bundle-based interconnect systems. Their analysis shows that the DWCNT bundle-based interconnect will be more suitable for the next generation of interconnect technology as compared to the SWCNT bundle counterpart.

Li et al. [52] reviewed the researches in CNT and GNR interconnects. They provided both electrical and thermal models of CNT and GNR interconnects. They showed that SWCNT, DWCNT, and MWCNT can provide better performance than copper interconnects. It is also shown that CNT can be applied as through-silicon vias (TSVs) in 3D integrated circuits.

Li and Banerjee [53] studied high-frequency effects in CNT interconnects. They developed a frequency-dependent impedance extraction method for both SWCNT- and MWCNT-based interconnects. They showed that CNT-based planar spiral inductors can achieve more than 3× higher Q factor than that of Cu-based counterparts without using any magnetic materials or Q-factor enhancement techniques.

Fathi and Forouzandeh [54] performed stability analysis in CNT interconnects based on transmission line modeling and using the Nyquist stability criterion.

Nasiri et al. [55] performed a similar stability analysis on GNR interconnects.

Sarto et al. [56] proposed hybrid transmission line-quantum mechanical models for the analysis of the signal propagation along metallic and quasi-metallic SWCNT and bundles of SWCNTs. The model is developed based on electron waveguide formalism in time and frequency domains, taking into account the damping effect produced by electron scattering. Sarto and Tamburrano [57] derived an equivalent single-conductor (ESC) model of a MWCNT interconnect. Sarto and Tamburrano [58] performed a comparative analysis on multilayer GNR (MLGNR) and MWCNT interconnects. They also studied the RF performance up to 100 GHz. D'Amore and coauthors [59] performed fast transient analysis in SWCNT bundle and MWCNT interconnects.

Pasricha and coauthors [60] evaluated the performance of CNT global interconnects at the system level. Their experimental results indicate that although SWCNTs are not as suitable for global interconnect buses, global MWCNT buses can provide performance speedups. Global interconnect buses implemented with SWCNT bundles and mixed bundles also lead to performance gains over copper global buses.

Naeemi and Meindl [61] proposed a compact physics-based circuit model for GNR interconnect. Xu et al. [62] derived the conductance model of GNR using tight-binding model and linear response Landauer formula. They also performed delay analysis in GNR interconnects and compared the results with other interconnect materials, such as copper, tungsten (W), and CNT.

Nasiri et al. [55] performed stability analysis in GNR interconnects. In their analysis, they studied the relative stability of MLGNR interconnects using the Nyquist stability criterion. Their analysis shows that when length and width are increased, MLGNR interconnects become more stable.

Lee et al. [63] analyzed the performance of monolithically integrated graphene interconnects on a prototype 0.35-μm CMOS chip. They have grown large graphene sheets that are further processed by standard lithography methods to produce narrow graphene wires up to 1 mm in length. They demonstrated end-to-end data communication on these graphene wires with bit-error-rate (BER) below 2×10^{-10}.

Sarkar et al. [64,65] presented a detailed methodology for the accurate evaluation of high-frequency impedance of graphene-based structures relevant to on-chip interconnects and low-loss on-chip inductors.

Yu et al. [66] investigated key reliability-limiting factors in bilayer graphene (BLG)/copper hybrid interconnect system. The results show that BLG displays an impressive current-carrying capacity ~100 times that of Cu.

Yu et al. [67] experimentally demonstrated stacked multilayer graphene (s-MLG) to be a new material system for high-performance carbon interconnects. They showed that the wire sheet resistance of s-MLG decreases with more layers stacking and shows better conductivity than that of the exfoliated MLG sample with the same layer numbers.

Lee et al. [68] fabricated graphene wires from large-area MLG sheets grown by CVD. The average thickness of multilayer graphene sheets was 10–20 nm with sheet resistances between 500 and 1000 Ω/sq.

Cui and coauthors [69] analyzed signal transmission characteristics in MLGNR interconnects for 14- and 22-nm technology nodes. An ESC model has been derived and transient response is studied in their work.

Rakheja and Naeemi [70] compared the performance and the energy dissipation of graphene spin interconnects in a nonlocal spin-torque (NLST) circuit against those of the CMOS circuit at the end of the silicon technology road map [71].

2.4 Works Related to CNT- and GNR-Based Field-Effect Transistors

Javey et al. [72] fabricated Schottky barrier CNTFET (SB-CNTFET) using nanotube metal junctions. They used semiconducting SWCNTs that showed room-temperature conductance near the ballistic transport limit of $4e^2/h$ and high current-carrying capability of ~25 mA per tube.

Durkop et al. [73] fabricated transistors using semiconducting CNTs with channel length exceeding 300 μm. They developed nanotubes in a tube furnace at 900°C.

Hoenlein et al. [74] discussed the applicability of CNTs in FETs. Zhou et al. [75] studied semiconducting SWCNTs in the diffusive transport regime.

Guo et al. [76] presented device physics of CNTFETs. They classified CNTFETs into two broad classes: (i) CNT MOSFET, which is similar to the traditional silicon MOSFET, and (ii) CNT MSDFETs, for metal source/drain FETs. For ballistic transport, they self-consistently solved the Poisson and Schrodinger equations using the nonequilibrium Green's function (NEGF) formalism.

The semiconducting CNTs are considered for the channel region of high-speed transistors due to their near ballistic electron transport. There are three different types of proposed CNTFET structures: SB-CNTFET, MOSFET-like CNTFET, and band-to-band tunneling CNTFET (BTBT-CNTFET). CNTFET has a structure that is similar to that of MOSFET. The ballistic transport operation and low OFF current makes CNTFET attractive for future nano-electronic circuits.

In the work of Raychowdhury et al. [77], it has been shown that MOSFET-like CNTFET is superior as compared to SB-CNTFET. The proposed MOSFET-like CNTFET [78–80] is likely to be scaled down to 10 nm channel length. The simulation program with integrated circuit emphasis (SPICE) compatible equivalent circuit of MOSFET-like CNTFET is proposed by Deng and Wong [79,80].

Sinha et al. [81] developed a noniterative physics-based compact model for CNT transistors and interconnects. They showed that for a SB-CNTFET with the diameter range of 11.5 nm, the circuit can be more than 8× faster than that of 22-nm CMOS, with the tolerance to the variation in contact materials.

Lin et al. [82] designed the static random access memory (SRAM) cell using MOSFET-like CNTFET. They used dual-chirality in the design. It is shown that the CNTFET-based SRAM cell tolerates the process, power supply voltage, and temperature variations significantly better than its CMOS counterpart.

Zhang et al. [83] presented a design technique to overcome two major sources of CNTFET imperfections: metallic CNTs and CNT density variations.

2.5 Practical Circuits

Close et al. [84] reported the fabrication of the first stand-alone integrated circuit combining silicon transistors and individual CNT interconnect wires on the same chip. The operating frequency reported is above 1 GHz. They assembled MWCNT interconnects on top of a CMOS chip fabricated in a 0.25-μm silicon CMOS process containing about 11,000 transistors. They used MWCNTs as electrical wires interconnecting various stages of conventional ring oscillators.

2.6 Summary

A detailed literature review is presented in this chapter. The CNT and GNR growth technologies are presented along with their merits and demerits. Next, the modeling works on CNT and GNR interconnects are discussed. Several analysis works are reviewed thoroughly for the applicability of CNT and GNR as nanointerconnects for future generation VLSI circuits.

3

Modeling of CNT and GNR Interconnects

3.1 Introduction

Integrated circuits have two major components: semiconducting devices, such as diodes, or transistors, and interconnects. Interconnects refer the physical connecting medium made of thin metal films between several electrical nodes in a semiconducting chip. The purpose of the interconnects is to transmit signals from one point to another without any distortion. Depending on the length of the interconnects, they are classified into three types: (1) global interconnect (length > 100 μm), (2) semiglobal or intermediate interconnect (10 μm > length > 100 μm), and (3) local interconnect (length < 10 μm).

There are different interconnect technologies.

- Metallic interconnects
- Optical interconnects
- Superconducting interconnects

As far as the metallic interconnect technology is concerned, aluminum was used as the interconnect material. However, due to lower resistivity of copper, the aluminum was subsequently replaced by copper.

As the very large-scale integration (VLSI) technology advances, the number of components integrated into a chip also increases aggressively. The devices and the interconnects are miniaturized to a large extent. In state-of-the-art VLSI chips, the devices are placed so compact that there is not enough space for the interconnects. Also due to the huge number of connections required in a chip, the interconnections are not possible in a single layer. Several layers of interconnects are required to complete all the required connections between the devices. In the multilayer interconnection structure, alternate layers of interconnects are drawn in a perpendicular direction. Low-k dielectric materials are used to isolate several metal layers. The vertical interconnection structure is known as via. Vias are used to make connections between alternate metal layers.

Typically, the cross-section of the interconnects increases from the bottom to the top metal layers. The local interconnects are formed using the bottom metal layers. The intermediate interconnects are formed using the middle metal layers, whereas the global interconnects are formed using the top metal layers. The cross-section of the wires is such that the vertical dimension is more than the lateral dimension. This is to accommodate more interconnects in a given area.

As the interconnect dimension is reduced, surface scattering and grain boundary scattering become prominent, and the bulk resistivity of the material is increased significantly. For example, the bulk resistivity of copper, which is 1.7 μΩ-cm, is increased 3–4 times in the subnanometer regime.

The other important problem is the susceptibility to electromigration. As the interconnect dimensions are scaled down, the cross-sectional area is reduced. At the same time, due to the huge integration of the devices, possibility of simultaneous switching also increases. Therefore, the current density of the interconnects increases significantly. When the current density increases beyond a threshold, the electron knocks some of the host metal atoms to migrate from their original location. This phenomenon is known as electromigration. Due to electromigration, the thin metal films may break, creating voids, or form hillocks creating short circuits.

The interconnect delay nowadays contributes to 50% of the total path delay. Apart from timing issues, there are signal integrity and device reliability issues due to the interconnects. In order to estimate the effects of the interconnects on circuit performance, an accurate circuit equivalent model is required to be developed for complex interconnects. The equivalent model consists of resistance, inductance, and capacitance (RLC). Resistance is mainly determined by the geometry of the interconnects only and does not depend on the distribution of the interconnects in its surroundings. On the contrary, capacitance and inductance are strongly affected by the geometry and the distribution of neighboring conductors.

After the invention of the carbon nanotube (CNT) and graphene, there are significant efforts put forward to use them as VLSI interconnects in the nanometer regime. Both CNT and graphene show excellent electrical, thermal, and mechanical properties. The mean free path (MFP) of CNT (λ_{CNT}) is several times larger than that of copper ($\lambda_{Cu} \approx 40$ nm at room temperature).

In this chapter, we discuss the electrical modeling of nanointerconnects using CNT and graphene.

3.2 Single-Walled CNT

The electrical equivalent circuit of CNT was proposed by Burke [1] based on Luttinger's liquid theory. Figure 3.1 shows the transmission line model of an isolated CNT. It is basically modeled as a 1D quantum wire. According to the Landauer–Buttiker formalism, the conductance of a 1D quantum wire with Γ channels in parallel is given by

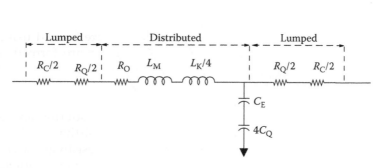

FIGURE 3.1
Transmission line model of an isolated CNT.

$$G = \frac{e^2}{h}\Gamma \times T \tag{3.1}$$

where:
  h is Planck's constant
  e is the electronic charge
  Γ is number of modes/channels
  T is transmission probability

An isolated CNT has four conducting channels ($\Gamma = 4$) in parallel. The spin degeneracy and sublattice degeneracy of electrons in graphene contribute four conducting channels.

Therefore, an isolated CNT has a quantum resistance, assuming perfect metal–CNT contact ($T = 1$), given by

$$R_Q = \frac{h}{4e^2} = 6.45\,\text{k}\Omega \tag{3.2}$$

This is the fundamental quantum resistance of an isolated CNT of length less than the electron MFP (λ_{CNT}). For CNTs with length $l < \lambda_{CNT}$, the transport is purely ballistic. However, for lengths $l > \lambda_{CNT}$, there is scattering-induced ohmic resistance, which is given by

$$R_{CNT} = R_C + R_Q + R_O = R_C + \frac{h}{4e^2}\left(1+\frac{l}{\lambda_{CNT}}\right) \tag{3.3}$$

The contact resistance and quantum resistance are independent of CNT length and are therefore modeled by two lumped values $R_C/2$ and $R_Q/2$ at both the ends.

In addition to the magnetic inductance (L_M) and electrostatic capacitance (C_E) of the traditional interconnect model, there are kinetic inductance (L_K) and quantum capacitance (C_Q). An isolated CNT of diameter d at height ht over a ground plane is shown in Figure 3.2.

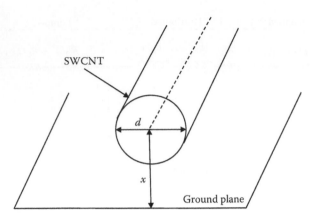

FIGURE 3.2
CNT over a ground plane.

The per-unit-length (p.u.l.) magnetic inductance is expressed as

$$L_M = \frac{\mu}{2\pi}\ln\left(\frac{x}{d}\right) = \frac{\mu}{2\pi}\cos h^{-1}\left(\frac{x}{d}\right) \tag{3.4}$$

where:
  x is the height of the CNT from ground plane
  d is the diameter of CNT

For $d = 1$ nm and $ht = 1$ μm, $L_M = 1.38$ pH/μm.
  The kinetic inductance is given by

$$L_K = \frac{h}{2e^2 v_F} \tag{3.5}$$

where v_F is the Fermi velocity, which is 8×10^5 m/s for CNT. The value of kinetic inductance has been numerically calculated to be 16 nH/μm. It is important to note that for an isolated CNT, magnetic inductance is insignificant and the kinetic inductance dominates. As there are four conducting channels in a CNT, the effective kinetic inductance of an isolated CNT is $L_K/4$.
  The electrostatic capacitance is given by

$$C_E = \frac{2\pi\varepsilon}{\ln(x/d)} \tag{3.6}$$

For $x = 1$ μm and $d = 1$ nm, electrostatic capacitance is calculated numerically as $C_E = 50$ aF/μm.

The expression for the quantum capacitance is given by

$$C_Q = \frac{2e^2}{hv_F} \qquad (3.7)$$

Numerically, the value of quantum capacitance is calculated as $C_Q = 96.8$ aF/μm. The effective quantum capacitance of an isolated CNT is $4C_Q$ due to four conducting channels.

3.3 Multiwalled CNT

MWCNT is a CNT structure with several concentric shells of rolled graphene sheets as shown in Figure 3.3.

The diameter of MWCNT varies from several nanometers to tens of nanometers. Although SWCNTs can be either metallic or semiconducting depending on their chirality, MWCNTs are always metallic. Moreover, MWCNTs have similar current-carrying capacity (as metallic SWCNTs) but are easier to fabricate than SWCNTs due to easier control of the growth process.

An isolated MWCNT on an infinite ground plane is shown in Figure 3.3. The separation between the nanotube center and the ground is x, the diameter of the outermost shell is D_{max}, the diameter of the innermost shell is D_{min}, and the interval between two adjacent shells is $\delta = 0.34$ nm, which is the van der Waals gap.

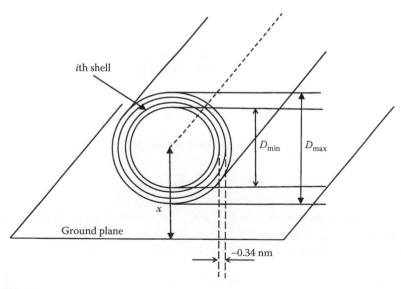

FIGURE 3.3
MWCNT over a ground plane.

The circuit model of MWCNTs is much more complex than the traditional RLC distributed interconnect model. Each shell in MWCNTs have different parameters and there are couplings between neighboring shells.

3.3.1 Modeling of Conducting Channels

The number of shells in MWCNT is diameter dependent [2,3], that is,

$$M_{shell} = 1 + \frac{D_{max} - D_{min}}{2\delta} \tag{3.8}$$

The number of conducting channels (considering spin degeneracy) in any shell is given by [2]

$$M_{shell}(D) = \alpha D + \beta \tag{3.9}$$

where:
  D is the diameter of the shell
  $\alpha = 0.0612 \text{ nm}^{-1}$
  $\beta = 0.425$

The ratio D_{min}/D_{max} is assumed to be approximately 1/2, in the modeling works [2] and [4]. The same is supported by the experimental work in [5]. Thus, the number of shells v of the MWCNT is determined by

$$v = 1 + \text{Integer}\left[\frac{(D_{max} - D_{max}/2)}{2\delta}\right] \tag{3.10}$$

where Integer[] indicates that only the integer part is taken into account. If we denote the shells from the outer to the inner as 1, 2, … , k, … v, the diameter of the kth shell is given by

$$D_k = D_{max} - 2\delta(k-1) \text{ for } 1 \le k \le v \tag{3.11}$$

where δ is the van der Waals gap and is equal to 0.34 nm. The innermost diameter in Figure 3.3 is

$$D_{min} = D_{max} - 2\delta(k-1) \tag{3.12}$$

Note that, although the ratio of D_{min}/D_{max} is assumed to be 1/2, D_{min} may be larger than $D_{max}/2$ because D_{max} may not be an integer multiple of δ. The number of conducting channels of the kth shell is given by

$$M_k = \alpha D_k + \beta \tag{3.13}$$

Hence, the total number of conducting channels (M) is given by the sum of the conducting channels (M_k) of all the shells:

$$M = \sum_{k=1}^{v} M_k \tag{3.14}$$

3.3.2 Resistance of Individual Shell

The resistance of a shell consists of three parts [4]: the quantum contact resistance R_Q, the scattering-induced ohmic resistance R_O, and the imperfect contact resistance R_C. R_O occurs only if the length of the nanotube is larger than the electron MFP. R_Q and R_O are intrinsic, and R_C is due to fabrication process.

The value of the intrinsic resistance of each shell is determined by

$$R_{\text{shell}}^k = R_Q^k + R_O^k = \frac{h}{2e^2 M_k}\left(1 + \frac{l}{\lambda_k}\right) \tag{3.15}$$

Or

$$R_{\text{shell}}^k = R_Q^k + R_O^k = \frac{h}{2e^2 M_k} + \frac{h}{2e^2 M_k} \cdot \frac{l}{\lambda_k} \tag{3.16}$$

where:
$h/(2e^2) = 12.9 \text{ k}\Omega$

l, λ_k, and M_k are the length, MFP, and number of conducting channels of kth shell in the MWCNT, respectively.

The imperfect contact resistance R_C can range from 0 to 100 kΩ for different growth processes. Recently, as demonstrated in [5] and [6] R_C in MWCNT could be very small compared to the total resistance.

3.3.3 Inductance of Individual Shell

The p.u.l. kinetic inductance of kth shell is given by

$$L_K^k = \frac{L_K}{2M_k} \tag{3.17}$$

The p.u.l. magnetic inductance of each shell is negligible ($\sim$pH/μm) as compared to kinetic inductance.

3.3.4 Capacitance of Individual Shell

The p.u.l. quantum capacitance of the kth shell is given by

$$C_Q^k = C_Q \times 2M_k \tag{3.18}$$

The p.u.l. electrostatic mutual capacitance between the kth and $(k + 1)$th shells is given by

$$C_C^{k,k+1} = \frac{2\pi\varepsilon_0}{\ln(D_{k+1}/D_k)} \tag{3.19}$$

The p.u.l. electrostatic capacitance of the outermost shell is given by

$$C_E = \frac{2\pi\varepsilon_0}{\ln(1.5D_{max}/D_{max})} \tag{3.20}$$

Only the outermost shell has electrostatic capacitance with ground, whereas the inner shells are shielded.

The equivalent circuit of an MWCNT having v number of shells is shown in Figure 3.4.

3.3.5 RLC of MWCNT

The total resistance can be calculated as the parallel combination of v number of resistances due to each shell and is given by

$$R_{MWCNT} = \left[\frac{1}{\sum_{k=1}^{v} 1/R_{shell}^k} \right] \tag{3.21}$$

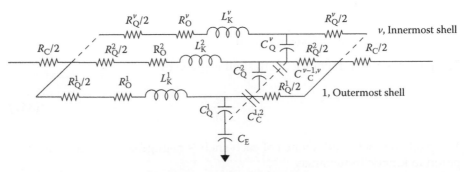

FIGURE 3.4
Equivalent circuit of an MWCNT having v number of shells.

The total kinetic inductance of all of the conducting shells in an MWCNT is [7]

$$L^K_{MWCNT} = \left[\frac{1}{\sum_{k=1}^{v} 1/L^k_K} \right] \tag{3.22}$$

The total quantum capacitance of all of the conducting shells in an MWCNT is [7]

$$C^Q_{MWCNT} = \sum_{k=1}^{v} C^k_Q \tag{3.23}$$

3.4 SWCNT Bundle

The interconnect wire is formed by a bundle of identical SWCNTs, as shown in Figure 3.5.

If w and t are the interconnect width and thickness, respectively, the number of CNTs along x and z direction can be expressed as

$$p_x = \frac{w - d}{y} \tag{3.24}$$

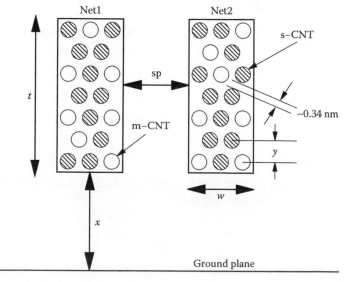

FIGURE 3.5
Interconnect structure using SWCNT bundle.

and

$$q_z = \frac{t-d}{\sqrt{3}/2y+1} \tag{3.25}$$

where d is the diameter of CNT.

For a densely packed bundle, the inter-CNT distance is $y = d + \delta$, where $\delta = 0.34$ nm (van der Waals gap). Using the expressions of p_x and q_z, the total number of CNTs in a bundle can be expressed as

$$n_{CNT} = p_x q_z - q_z/2, \quad \text{if } q_z \text{ is even} \tag{3.26}$$

$$n_{CNT} = p_x q_z - (q_z - 1)/2, \quad \text{if } q_z \text{ is odd} \tag{3.27}$$

Due to the lack of control on chirality, a bundle consists of both metallic and semiconducting nanotubes. Statistically, it has been found that *metallic fraction P_m* is one-third in a bundle, that is, one-third of the nanotubes are metallic and the rest are semiconducting. Only the metallic CNTs contribute to current conduction, and hence, the effective number of conducting channels in a bundle with both metallic and semiconducting CNTs is less than that of a bundle with all metallic CNTs. A bundle with all metallic CNTs is termed as *densely packed* bundle, and a bundle with both metallic and semiconducting CNTs is termed as *sparsely packed* bundle.

As the sparsely packed bundle may have metallic CNTs surrounded by semiconducting CNTs, the sparsely packed bundle can be modeled considering larger inter-CNT spacing, that is, $y > d + \delta$.

3.4.1 Resistance of a Bundle

The resistance of a SWCNT bundle consisting of n_{CNT} number of SWCNTs is expressed as [8]

$$R_b = \frac{R_{CNT}}{P_m n_{CNT}} \tag{3.28}$$

where P_m is the fraction of metallic tubes in a bundle.

3.4.2 Inductance of a Bundle

The CNTs have both kinetic and magnetic inductances. The kinetic inductance of the bundle is expressed as [9]

$$L_b = l\frac{L_K/4}{P_m n_{CNT}} \tag{3.29}$$

In [10], Srivastava et al. calculated the magnetic inductance of the CNT bundle by dividing the magnetic inductance of an isolated CNT by the number of CNTs which underestimates the magnetic inductance. The mutual inductance between the CNTs in a bundle is neglected in [10]. The magnetic inductance can be calculated accurately following the modeling technique proposed by Nieuwoudt and Massoud [9]. In [9], the magnetic inductance of the CNT bundle is calculated considering the mutual inductances between the CNTs in a bundle using the partial inductance modeling approach based on the partial element equivalent circuit (PEEC) method. The partial self-inductance (L_m) of a single SWCNT [9] is given by

$$L_m = \frac{\mu_0 l}{2\pi} \left[\ln \frac{l}{d} + \frac{1}{2} + \frac{2d}{3l} \right] \tag{3.30}$$

where:
  l is the length
  d is the diameter of the SWCNT

The mutual inductance between (M_m) between two parallel SWCNTs of equal length [9] is given by

$$M_m = \frac{\mu_0 l}{2\pi} \left[\ln \left(\rho + \sqrt{1+\rho^2} - \sqrt{1 + \frac{1}{\rho^2}} + \frac{1}{\rho} \right) \right] \tag{3.31}$$

where $\rho = l/y$ and y = center-to-center spacing between two SWCNTs (see Figure 3.5). According to the PEEC method, the overall loop inductance of the bundle is given by

$$L_b^M = i_n^T L_{mat} i_n \tag{3.32}$$

where:
  i_n is a vector with normalized current in each SWCNT
  L_{mat} is the partial inductance matrix [9] given by

$$L_{mat} = \begin{bmatrix} L_m^1 & M_m^{1,2} & \cdots & M_m^{1,n} \\ M_m^{2,1} & L_m^2 & \cdots & M_m^{2,n} \\ \vdots & \vdots & \ddots & \vdots \\ M_m^{n,1} & M_m^{n,2} & \cdots & L_m^n \end{bmatrix} \tag{3.33}$$

where:
  L_m^j is the partial self-inductance of the jth SWCNT
  $M_m^{i,j}$ is the partial mutual inductance between the ith and jth SWCNTs

Hence, effective inductance of a SWCNT bundle is the series combination of kinetic and magnetic inductances and can be written as

$$L_b = L_b^K + L_b^M \tag{3.34}$$

3.4.3 Capacitance of a Bundle

The quantum capacitance of a CNT bundle is given by

$$C_b^Q = 4C_Q(P_m n_{CNT})l \tag{3.35}$$

The electrostatic capacitance of a SWCNT bundle is almost equal to that of copper wire of same cross-sectional dimensions [10]. It has also been found that the metallic fraction (P_m) of CNTs in a bundle does not affect the electrostatic capacitance significantly. In [11], it is reported that the capacitance of a sparsely packed bundle with four CNTs at the corners is just 20% less than that of a densely packed bundle. Therefore, it can be assumed that the electrostatic capacitance of a SWCNT bundle is same as that of copper wire of identical cross-section. Thus, the effective capacitance of SWCNT bundle is the series combination of its electrostatic capacitance and quantum capacitance.

3.5 MWCNT Bundle

Two different structures are proposed for interconnect using MWCNT bundle as shown in Figure 3.6. These structures are suitable only when the aspect ratio is two. Otherwise, smaller diameter MWCNTs must be used to fit the MWCNTs within the specified interconnect cross-section.

The resistance, kinetic inductance, and quantum capacitance of a MWCNT bundle is R_{MWCNT}/n_{CNT}, L_{MWCNT}^K/n_{CNT}, and $n_{CNT} \times C_{MWCNT}^Q$, where n_{CNT} is the number of MWCNTs in a bundle. The magnetic inductance of an MWCNT bundle is calculated using the methodology described for SWCNT bundles. In [10] and [11], the authors reported that the electrostatic coupling and ground capacitances of CNT bundle of smaller diameter CNTs are same as that of copper wire of equal dimension. Hence, it can be assumed that

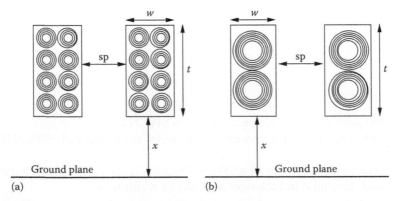

FIGURE 3.6
Interconnect structure using MWCNT bundle.

electrostatic and coupling capacitance for the smaller diameter MWCNT bundle-based interconnect (Figure 3.6a) are same as that of copper wire. For MWCNT bundles with large diameter MWCNTs (Figure 3.6b), the capacitance can be calculated using the model described in [7].

3.6 Modeling of Graphene Nanoribbon

In graphene, the carbon atoms are arranged in a honeycomb structure, as shown in Figure 3.7. Though GNR and CNT are derived from the basic graphene structure, there are differences. The difference arises due to the boundary conditions. In CNT, the wave function is periodic along the circumference. In GNR, the wave function vanishes at the boundary of the two edges.

Depending on the orientation of carbon atoms, the edge of the graphene sheet is either armchair or zigzag. Zigzag GNR is always metallic, whereas armchair GNR can be either semiconducting or metallic depending on the number of carbon rings across the width. For interconnect applications, zigzag GNR is proposed due to its metallic property.

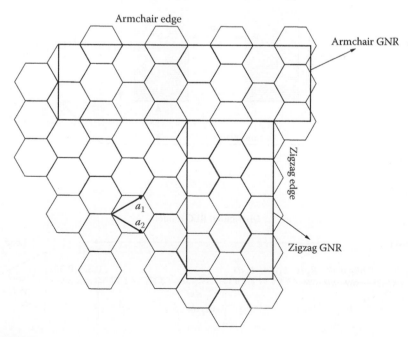

FIGURE 3.7
Lattice structure of graphene sheet.

Due to the high resistance of monolayer GNR, a multilayer GNR structure is proposed for the interconnects [12,13]. A multilayer GNR structure is modeled for VLSI interconnect as shown in Figure 3.8.

In Figure 3.10, the interconnect thickness is t, width is w, height from ground plane is x, and spacing between the interconnects is sp. The spacing between each graphene layer is δ (= 0.34 nm), which is the van der Waals gap [12]. The interconnect is assumed to have dimensions according to the specifications in International Technology Roadmap for Semiconductors (ITRS) [14] for 16-nm technology node. The number of graphene layers is given by

$$M_{\text{layer}} = 1 + \text{Integer}\left[\frac{t}{\delta}\right] \qquad (3.36)$$

Figure 3.9 shows the schematic diagram of an interconnect model [15] with a driver and a load at both ends of the interconnect. In Figure 3.9, the interconnect is modeled by a distributed RLC network, where R_C is the resistance due to imperfect contacts, R_Q is the quantum resistance, and R_S (= R_Q/λ) is the scattering resistance p.u.l., where λ is the MFP of electron in GNR. The quantum resistance is defined as [12]

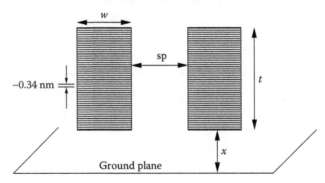

FIGURE 3.8
Multilayer GNR structure.

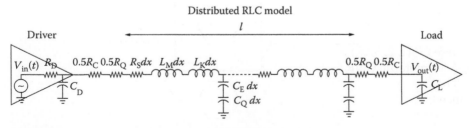

FIGURE 3.9
Schematic of the RLC interconnect model of GNR interconnect.

$$R_Q = \frac{h/2e^2}{M_{ch}M_{layer}} = \frac{12.94 \text{ k}\Omega}{M_{ch}M_{layer}} \tag{3.37}$$

where:
  M_{ch} is the number of conducting channels (modes) in one layer
  M_{layer} is the number of GNR layers
  h is the Planck's constant (= 6.626×10^{-34} J-s)
  e is the electronic charge (= 1.6×10^{-19} C)

It is assumed that λ is equal to 1 μm [16].
  The p.u.l. quantum capacitance [12] is expressed as

$$C_Q = M_{ch}M_{layer}\frac{4e^2}{hv_F} = M_{ch}M_{layer} \times 193.18 \text{ aF/μm} \tag{3.38}$$

where v_F is the Fermi velocity = 8×10^5 m/s for GNR [16]. The p.u.l. kinetic inductance [12] is expressed as

$$L_K = \frac{h/4e^2v_F}{M_{ch}M_{layer}} = \frac{8.0884}{M_{ch}M_{layer}} \text{nH/μm} \tag{3.39}$$

The number of conducting channels in monolayer graphene [12,15] is given by

$$M_{ch} = \sum_{j=1}^{n_C}\left[1+e^{(\varepsilon_{j,n}-\varepsilon_F)/k_BT}\right]^{-1} + \sum_{j=1}^{n_V}\left[1+e^{(\varepsilon_F+\varepsilon_{j,h})/k_BT}\right]^{-1} \tag{3.40}$$

where:
  j (= 1, 2, 3,...) is a positive integer
  ε_F is Fermi energy
  k_B is the Boltzmann's constant
  T is absolute temperature
  n_C and n_V are the number of conduction and valence subbands, respectively
  $\varepsilon_{j,n}$ and $\varepsilon_{j,h}$ are the energies of electron and hole in the jth subband, as given
    by [16]

$$\varepsilon_j = \frac{hv_F}{2w}\left|j+\frac{1}{2}\right| \tag{3.41}$$

where j = (1, 2, 3, ...) is an integer. The number of conducting channels (M_{ch}) is 6 for metallic GNR of width 16 nm [17] for ε_F = 0.3 eV.
  The expressions for p.u.l. magnetic inductance and electrostatic capacitance [13,15] are given by

$$L_m = \frac{\mu \times x}{w} \tag{3.42}$$

and

$$C_e = \frac{\varepsilon \times w}{x} \tag{3.43}$$

However, in [12], the magnetic inductance and electrostatic capacitance are modeled using the predictive technology model [18]. The magnetic inductance and electrostatic and coupling capacitances of GNR are assumed to be same as that of a copper with equal dimensions.

3.7 Modeling of Copper Interconnects

3.7.1 Resistance of Copper Interconnects

At nanometer dimensions, the electrical resistivity (ρ) of copper wire is determined by two phenomena: surface scattering and grain boundary scattering. The surface scattering-based resistivity model was proposed by Fuchs [19] and Sondheimer [20], which is given by

$$\frac{\rho_{FS}}{\rho_0} = 1 + \frac{3}{4} \frac{\lambda_0}{w} (1 - z) \tag{3.44}$$

where:
ρ_0 is the resistivity of the bulk material
w is width of the wire
λ_0 is the MFP of the conduction electrons
z ($= 0.6$) is the Fuchs scattering parameter

The grain boundary-based resistivity model was proposed by Mayadas and Shatzkes (M–S) [21], which is given by

$$\frac{\rho_{MS}}{\rho_0} = \left[1 - \frac{3}{2}\beta + 3\beta^2 - 3\beta^3 \ln\left(1 + \frac{1}{\beta}\right) \right]^{-1} \tag{3.45}$$

where:

$$\beta = \frac{\lambda_0}{D} \frac{\Theta}{1 - \Theta} \tag{3.46}$$

Here D is the mean grain size and Θ is the reflection coefficient in the grain boundaries with values between 0 and 1. It is assumed that the mean grain size is same as the film width and $\Theta = 0.33$.

The resistance of the copper nanointerconnect can be modeled by combining the effects of surface and grain boundary scattering as

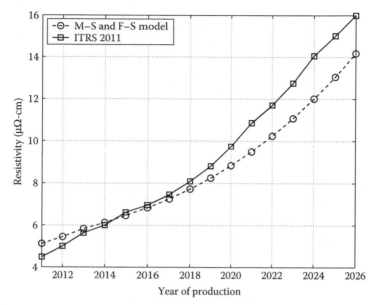

FIGURE 3.10
Resistivity of copper interconnect versus interconnect width for different technology nodes.

$$R_{Cu} = \rho_{Cu} \frac{l}{w \times t} = (\rho_{FS} + \rho_{MS}) \frac{l}{w \times t} \qquad (3.47)$$

where l, w, and t are the length, width, and thickness of the interconnect, respectively. Figure 3.10 illustrates the resistivity of the copper nanointerconnect as a function of wire width for sub-90-nm technology nodes. It is observed that the resistivity increases significantly many times over its bulk value of 1.7 μΩ-cm.

3.7.2 Inductance of Copper Interconnects

The self- and mutual inductances of copper nanointerconnects [18,22] are determined using the following standard formulae

$$L_S = \frac{\mu_0 l}{2\pi} \left[\ln\left(\frac{2l}{w+t} \right) + \frac{1}{2} + \frac{0.22(w+t)}{l} \right] \qquad (3.48)$$

and

$$M = \frac{\mu_0 l}{2\pi} \left[\ln\left(\frac{2l}{d_s} \right) - 1 + \frac{d_s}{l} \right] \qquad (3.49)$$

where:
  l is length
  w is width

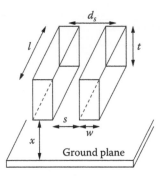

FIGURE 3.11
Schematic of a parallel interconnect structure over a ground plane.

t is thickness
μ_0 is permeability
d_s is pitch of the wires (see Figure 3.11)

3.7.3 Capacitance of Copper Interconnects

The capacitance with respect to ground plane and coupling capacitance between two adjacent lines [18] are given by

$$C_G = \varepsilon\left[\frac{w}{x} + 2.22\left(\frac{s}{s+0.7x}\right)^{3.19} + 1.17\left(\frac{s}{s+1.51x}\right)^{0.76} \times \left(\frac{t}{t+4.53x}\right)^{0.12}\right] \quad (3.50)$$

and

$$C_C = \varepsilon\left[\begin{array}{c} 1.14\dfrac{t}{s}\left(\dfrac{x}{x+2.06s}\right)^{0.09} + 0.74\left(\dfrac{w}{w+1.59s}\right)^{1.14} + \\[2mm] 1.16\left(\dfrac{w}{w+1.87s}\right)^{0.16} \times \left(\dfrac{x}{x+0.98s}\right)^{1.18} \end{array}\right] \quad (3.51)$$

where:
   ε is dielectric constant
   s is spacing between the interconnects (see Figure 3.11)
   x is height above ground plane

3.8 Interconnect Parameters

The interconnect parameters obtained from ITRS [14] are shown in Table 3.1.
   The RLC of the SWCNT bundle and MWCNT and GNR-based interconnects are shown in Tables 3.2 through 3.9.

TABLE 3.1

Technology Parameters from ITRS Road map

Year	2010	2013	2016	2019
DRAM 1/2 pitch (nm)	**45**	**32**	**22**	**16**
MPU M1 1/2 pitch (nm)	45	32	22	16
MPU gate length (nm)	18	13	9	6
No. of metal level (nm)	12	13	13	14
M1 wiring pitch (nm)	90	64	44	32
	90	64	44	32
	135	96	66	48
M1 A/R	1.8	1.9	2	2
	1.8/1.6	1.9/1.7	2.0/1.8	2.0/1.8
	2.4/2.2	2.5/2.3	2.6/2.4	2.8/2.5
Effective resistivity	4.08	4.83	6.01	7.34
($\mu\Omega$-cm)	4.08	4.83	6.01	7.34
	3.1	3.52	4.2	4.93
V_{DD} (V)	1.0	0.9	0.8	0.7
T_{ox} (nm)	1.0	0.9	0.8	0.7
Dielectric constant	2.5–2.8	2.1–2.4	1.9–2.2	1.6–1.9

TABLE 3.2

RLC Parameters of Copper Interconnects for Different Technology Nodes

Circuit Parameters	Length (μm)	45 nm	32 nm	22 nm	16 nm
Resistance (kΩ)	1	0.011	0.025	0.062	0.143
	5	0.056	0.124	0.31	0.717
	10	0.11	0.25	0.62	1.43
	50	0.56	1.24	3.1	7.17
	100	1.12	2.48	6.21	14.34
Inductance (pH)	1	0.66	0.72	0.79	0.85
	5	4.88	5.18	5.52	5.84
	10	11.14	11.75	12.43	13.07
	50	71.77	74.83	78.24	81.42
	100	157.4	163.52	170.33	176.7
Capacitance to ground (fF)	1	0.027	0.022	0.019	0.016
	5	0.133	0.108	0.094	0.081
	10	0.266	0.216	0.188	0.163
	50	1.328	1.080	0.941	0.813
	100	2.655	2.150	1.883	1.626

(Continued)

TABLE 3.2

(Continued) RLC Parameters of Copper Interconnects for Different Technology Nodes

Circuit Parameters	Length (µm)	45 nm	32 nm	22 nm	16 nm
Coupling capacitance (fF)	1	0.068	0.061	0.058	0.050
	5	0.341	0.306	0.292	0.252
	10	0.682	0.611	0.584	0.505
	50	3.411	3.056	2.922	2.523
	100	6.821	6.112	5.844	5.047

TABLE 3.3

Resistance of SWCNT Bundle-Based Interconnects at Room Temperature (300 K)

	Technology Node			
Length (µm)	45 nm	32 nm	22 nm	16 nm
Perfect contact ($R_C = 0$), *densely packed* ($P_m = 1$)				
1	0.0112	0.0197	0.0378	0.0679
5	0.0161	0.0288	0.0565	0.1041
10	0.0217	0.0393	0.0783	0.1468
50	0.0655	0.1223	0.2512	0.4862
100	0.1202	0.226	0.4671	0.9102
$R_C = 100$ *k*Ω, *densely packed* ($P_m = 1$)				
1	0.0359	0.0665	0.1353	0.2595
5	0.0408	0.0757	0.1541	0.2956
10	0.0464	0.0862	0.1759	0.3383
50	0.0902	0.1692	0.3488	0.6777
100	0.1449	0.2728	0.5647	1.1017
$R_C = 100$ *k*Ω, *sparsely packed* ($P_m = 1/3$)				
1	0.1076	0.1995	0.406	0.7784
5	0.1225	0.227	0.4623	0.8869
10	0.1391	0.2585	0.5276	1.015
50	0.2706	0.5075	1.0463	2.0332
100	0.4347	0.8185	1.6941	3.3052

TABLE 3.4

Inductance and Capacitance of SWCNT Bundle-Based Interconnects (Length = 10 μm)

Technology Node	45 nm	32 nm	22 nm	16 nm
Parameter	*Densely packed* ($P_m = 1$)			
L (nH)	0.010	0.0189	0.0395	0.0775
C (fF)	30.36	21.08	15.20	10.46
	Sparsely packed ($P_m = 1/3$)			
L (nH)	0.0301	0.0582	0.1200	0.2357
C (fF)	24.29	6.86	12.16	8.37

TABLE 3.5

Resistance, Inductance, and Capacitance of MWCNT-Based Interconnects for 16-nm Technology Node

Length (μm)	Resistance (kΩ)	Inductance (nH)	Capacitance (fF)		
			Quantum	Electrostatic	Coupling
Single MWCNT					
1	0.936	0.622	2.5113	0.0214	0.0229
5	1.225	3.111	12.5565	0.1072	0.1147
10	1.583	6.222	25.113	0.2144	0.2295
50	4.408	31.11	125.565	1.0720	1.1474
100	7.926	62.22	251.13	2.1439	2.2947
Double MWCNTs					
1	0.34	0.14	10.82	0.0168	0.0464
5	0.54	0.72	54.09	0.084	0.232
10	0.78	1.44	108.18	0.168	0.464
50	2.70	7.22	540.88	0.84	2.32
100	5.10	14.44	1081.76	1.68	4.64
MWCNT bundle					
1	0.47	0.31	5.0226	0.0213	0.0346
5	0.61	1.56	25.113	0.1064	0.1731
10	0.79	3.11	50.226	0.2128	0.3462
50	2.20	15.56	251.13	1.0642	1.7311
100	3.96	31.11	502.26	2.1285	3.4623

TABLE 3.6

Resistance, Inductance, and Capacitance of MWCNT-Based Interconnects for 22-nm Technology Node

Length (μm)	Resistance (kΩ)	Inductance (nH)	Capacitance (fF)		
			Quantum	Electrostatic	Coupling
Single MWCNT					
1	0.562	0.337	4.6363	0.0248	0.0266
5	0.689	1.685	23.1815	0.124	0.133
10	0.848	3.37	46.363	0.248	0.266
50	2.098	16.85	231.815	1.24	1.33
100	3.654	33.70	463.63	2.48	2.66
Double MWCNTs					
1	0.217	0.1123	13.9088	0.0194	0.0537
5	0.31	0.5617	69.5	0.097	0.2685
10	0.425	1.123	139.1	0.194	0.537
50	1.337	5.617	695.4	0.97	2.685
100	2.475	11.234	1390.9	1.94	5.37
MWCNT bundle					
1	0.281	0.169	9.273	0.0246	0.0401
5	0.345	0.843	46.363	0.123	0.2005
10	0.424	1.685	92.726	0.246	0.401
50	1.049	8.425	463.630	1.23	2.005
100	1.827	16.850	927.260	2.46	4.01

TABLE 3.7

Resistance, Inductance, and Capacitance of MWCNT-Based Interconnects for 32-nm Technology Node

Length (μm)	Resistance (kΩ)	Inductance (nH)	Capacitance (fF)		
			Quantum	Electrostatic	Coupling
Single MWCNT					
1	0.286	0.17	9.0794	0.0271	0.029
5	0.333	0.86	45.397	0.1354	0.1449
10	0.39	1.72	90.794	0.2708	0.2899
50	0.84	8.60	453.97	1.3541	1.4493
100	1.399	17.2	907.94	2.7081	2.8986
Double MWCNTs					
1	0.117	0.078	20.09	0.0212	0.0586
5	0.153	0.389	100.45	0.106	0.293

(Continued)

TABLE 3.7

(Continued) Resistance, Inductance, and Capacitance of MWCNT-Based
Interconnects for 32-nm Technology Node

Length (μm)	Resistance (kΩ)	Inductance (nH)	Capacitance (fF)		
			Quantum	Electrostatic	Coupling
10	0.198	0.778	200.90	0.212	0.586
50	0.551	3.889	1004.52	1.06	2.93
100	0.991	7.778	2009.04	2.12	5.86
MWCNT bundle					
1	0.143	0.085	18.16	0.0269	0.0437
5	0.167	0.430	90.79	0.1344	0.2187
10	0.195	0.860	181.59	0.2689	0.4373
50	0.420	4.300	907.94	1.3443	2.1867
100	0.700	8.600	1815.88	2.6886	4.3734

TABLE 3.8

Resistance, Inductance, and Capacitance of MWCNT-Based Interconnects for
45-nm Technology Node

Length (μm)	Resistance (kΩ)	Inductance (nH)	Capacitance (fF)		
			Quantum	Electrostatic	Coupling
Single MWCNT					
1	0.157	0.095	16.42	0.0316	0.0338
5	0.175	0.476	82.1	0.158	0.169
10	0.198	0.952	164.2	0.316	0.338
50	0.375	4.759	821	1.58	1.69
100	0.595	9.517	1642	3.16	3.38
Double MWCNTs					
1	0.070	0.042	37.09	0.0247	0.0683
5	0.086	0.211	185.452	0.1235	0.3415
10	0.106	0.421	370.904	0.247	0.683
50	0.262	2.106	1854.52	1.235	3.415
100	0.457	4.213	3709.04	2.47	6.83
MWCNT bundle					
1	0.079	0.048	32.84	0.0314	0.051
5	0.088	0.238	164.2	0.157	0.255
10	0.099	0.476	328.4	0.314	0.51
50	0.188	2.380	1642	1.57	2.55
100	0.298	4.759	3284	3.14	5.1

TABLE 3.9

RLC Parameters of GNR-Based Interconnects for 16-nm Technology Node

Length (μm)	Resistance (kΩ)	Inductance		Capacitance	
		L_k (nH)	L_m (pH)	C_Q (pF)	C_E (fF)
1	0.0454	0.0142	0.85	0.11	0.017
5	0.1362	0.0710	5.84	0.55	0.084
10	0.2497	0.1419	13.07	1.10	0.167
50	1.1579	0.7095	81.42	5.50	0.837
100	2.2931	1.419	176.7	11.01	1.673

3.9 Summary

In this chapter, we presented the equivalent circuit models for SWCNT, MWCNT, bundle of SWCNTs and MWCNTs, GNR, and the traditional copper-based interconnects. The interconnect parameters are obtained from ITRS for 45-, 32-, 22-, and 16-nm technology nodes. The RLC parameters of different interconnect systems can be computed for different technology nodes. The RLC values of the interconnects are used for further analyses presented in Chapters 4 through 7.

4

Timing Analysis in CNT Interconnects

4.1 Introduction

As the dimensions of copper wire decrease, its resistivity increases significantly due to surface roughness and grain boundary scattering [1]. The significant increase in interconnect resistance will lead to an increase in interconnect delay and reliability problems [2–4], which will impose limitations on both performance and reliability of very large-scale integration (VLSI) circuits. Carbon nanotubes (CNTs) have been proposed as a possible replacement for copper-based interconnects in future technology nodes [5–11].

An isolated CNT has very high quantum resistance of approximately 6.45 kΩ, which is independent of the length of the nanotube. Therefore, a bundle of CNTs connected in parallel is proposed to reduce the delay associated with the CNT-based interconnects [12,13].

With the advancement of process technology, the interconnect dimensions are scaled down from micrometer to nanometer dimensions. At the nanometer level, the variations in interconnect dimensions due to process variation have become a serious concern. The manufacturing process associated with CNT-based interconnects faces a challenge: controlling the CNT diameter and spacing as reported in [14–16]. The impact of the variation in CNT diameter and spacing on the interconnect delay is crucial in evaluating the timing characteristics of future CNT-based VLSI interconnects.

In this chapter, we discuss the analysis of the electrical properties of single-walled CNT (SWCNT) bundle with respect to three important circuit-operating conditions: process, temperature, and voltage (PTV). A review of related works already published is described in the following text. In [17], Narasimhamurthy and Paily investigated the impact of bias voltage on the magnetic inductance of the CNT interconnects. Naeemi and Meindl [18] analyzed the effect of large bias voltage on the mean free path (MFP) and hence on the conductance of CNT interconnects. They found that the effect of backscattering due to optical and zone-boundary phonons on the conductance is negligible. Budnik et al. [19] modeled voltage-dependent resistance of CNT bundle-based interconnects. The temperature-dependent model of CNT resistance is analyzed in [20–22]. The same temperature-dependent model has been used in [23–26] for electrothermal characterization of CNT interconnects. Sun and Luo [27] analyzed crosstalk using a simplistic process variation considering a linear approximation method. In [28], Alam et al. modeled process variation considering variations in interconnect width and height without considering a variation in CNT parameters. Nieuwoudt and Massoud [29] modeled process variation considering several parameter variations in CNT bundle interconnects.

In this chapter, we present the electrical equivalent circuit parameters at different PTV conditions. Under different PTV conditions, the CNT circuit parameters are extracted and the equivalent circuit is developed. The SPICE simulations are then performed to calculate the delay through the SWCNT bundle for different PTV conditions. A comparison is made with copper-based interconnects for 32- and 16-nm International Technology Roadmap for Semiconductors (ITRS) technology nodes.

4.2 CNT Model with PTV Variations

In this section, we present the model for SWCNT bundle with PTV variations.

4.2.1 Modeling Process Variation

CNTs are synthesized using several growth techniques such as arc-discharge, chemical vapor deposition (CVD), laser ablation, and template-assisted growth [14], out of which the CVD method is the most commercially viable technique. To form a VLSI interconnect, the CNTs must be bundled in order to reduce their large intrinsic resistance. Controlling the diameter of nanotubes in a bundle represents one of the most challenging issues in developing nanotube growth methods. When CNTs are grown in a bundle, their diameter often varies. It has been found in [9,10,15,16,30] that the diameter variation follows the Gaussian distribution. The range of diameter that is fabricated is from 0.8 to 1.5 nm. The shift in the mean nanotube diameter is approximately 0.2 nm (1.27–1.47 nm [16]) with full-width at half-maxima of ~0.2 nm. In [10], Lu et al. reported that CNT diameter follows Gaussian fits with mean diameter 0.91, 1.07, and 1.78 nm and standard deviations are 0.18, 1.12, and 1.02 nm, respectively. Another important process variation is one that leads to variation in metallic percentage. Typically, one-third ($P_m = 1/3$) of the CNTs are metallic in a bundle [31]. The remaining two-thirds are semiconducting and do not contribute to any current conduction.

It has been also found that the number of conducting channels also is a function of CNT diameter [22]. In CNTs with small diameter (<3 nm), increasing the temperature or diameter does not change the number of conducting channels. However, in large-diameter nanotubes, increasing the diameter or temperature linearly increases the number of conducting channels [22]. Assuming that one-third of the shells are metallic, the average number of channels for a shell is expressed as

$$m(d) = \begin{cases} \alpha Td + \beta & \text{for } d > d_T/T \\ 2/3 & d \le d_T/T \end{cases} \tag{4.1}$$

where:
  α is $2.04 \times 10^{-4}\,\text{nm}^{-1}\text{K}^{-1}$
  β is 0.425
  d_T is 1300 nm-K

For instance, at $T = 300$ K, the average number of channels becomes proportional to the diameter for $d > 4.3$ nm. In our analysis, the CNT diameter is within 1.5 nm, and hence, the average number of channels is 2/3 CNTs in a bundle.

In [32–34], Das et al. analyzed the effect of process variation on the resistance, inductance, and capacitance (RLC) equivalent model parameters of CNT nanointerconnects. They assumed nanointerconnects formed by bundles of SWCNT. As the number of metallic CNTs in a bundle determines the number of conducting channels in the nanointerconnect, it has a significant impact on the electrical performance of CNT bundles. Therefore, due to the process variation, the diameter is varied, and hence the equivalent RLC parameters change for a given width and thickness of the nanointerconnect.

Process variations can lead to three types of parameter variation in a CNT bundle: (1) CNT diameter, (2) spacing (inter-CNT distance) (x in Figure 4.1), and (3) height from the ground plane. Table 4.1 illustrates the process variations in CNT bundles in the works [32–34]. The distribution of CNTs $m(d)$ in a bundle is calculated using Gaussian distribution using Equation 4.2 with the mean (d_μ) values of CNT diameter of 1.1, 1.2, and 1.3 nm, respectively. The standard deviation (d_σ) is taken as 0.1 and 0.2 nm.

$$m(d) = \frac{1}{d_\sigma \sqrt{2\pi}} \exp\left[-\frac{1}{2}\left(\frac{d - d_\mu}{d_\sigma} \right)^2 \right] \tag{4.2}$$

In [32–34], the authors investigated the impact of only CNT parameter. They did not consider the variation in height from the ground plane. Four different values of spacing, $0d$, $0.5d$, $1d$, and $1.7d$, are considered in our analysis. These four values of spacing are used to model different metallic fractions ($P_m = 1, 2/3, 1/2, 1/3$) of CNTs in a bundle.

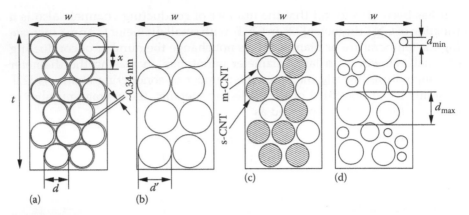

FIGURE 4.1

Illustration of SWCNT bundles: The black circle represents the CNT circumference. The outer gray circle has a radius of $(d + \delta)/2$. It is used to represent the intertube spacing (δ). (a) densely packed bundle with SWCNTs of same diameter d, (b) densely packed bundle with SWCNTs of same diameter d' (where $d' > d$), (c) sparsely packed bundle with SWCNTs of same diameter d, (d) densely packed bundle with SWCNTs of different diameter ($d_{min} \leq d \leq d_{max}$) with mean d_{μ}.

TABLE 4.1

Process Variation in CNT Bundle

Process Corner	Mean Diameter (d_μ)	Standard Deviation (d_σ)	Spacing	Inter-CNT Distance (x in Figure 4.1)	Packing Density
Min	1.1 nm	0.1 nm	$0d$	$d + \delta$	Dense
Nom	1.2 nm	0.2 nm	$0.5d$	$1.5d + \delta$	Sparse
Max	1.3 nm		$1d$	$2d + \delta$	
			$1.7d$	$2.7d + \delta$	

Figures 4.2 and 4.3 show the CNT distribution in CNT bundle of width 16 nm and thickness 32 nm (corresponding to 16-nm technology node [5]) with different values of CNT diameter and spacing. We considered in this work two ITRS technology nodes 32 and 16 nm. CNTs in a bundle with different mean diameter, spacing, and standard deviation are shown in Table 4.2 for 32-nm technology nodes and in Table 4.3 for 16-nm technology nodes, respectively.

From Tables 4.2 and 4.3, it is found that the sigma variation does not have any significant effect on the number of CNTs in a bundle. However, spacing variation has a large impact on the number of CNTs in a bundle.

It can be also found that a spacing of $1.7d$ corresponds to a sparse bundle with metallic fraction of 1/3, which is the value of metallic fraction statistically found. The spacing $0d$ corresponds to a densely packed bundle. The spacing $0.5d$ and $1d$ correspond to metallic fractions of 2/3 and 1/2, respectively.

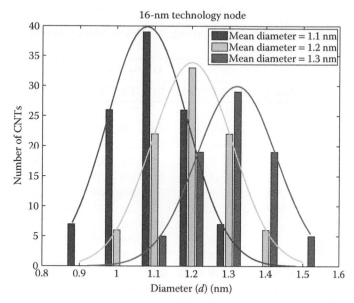

FIGURE 4.2
(See color insert.) Distribution of CNTs in a bundle of width 16 nm and thickness 32 nm. The interconnect dimensions are as per 16-nm technology node. The bundle is assumed to be densely packed, that is, with zero spacing.

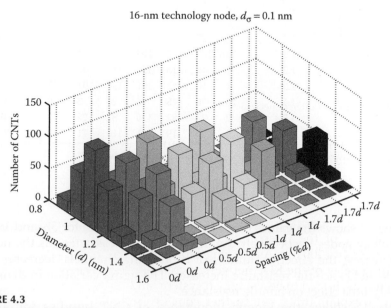

FIGURE 4.3
(See color insert.) CNT distribution in a bundle with different CNT diameter and spacing. The spacing is expressed as % of CNT diameter (d) with four different values ($0d$, $0.5d$, $1d$, and $1.7d$). Bundle width, $w = 16$ nm, and thickness, $t = 32$ nm.

TABLE 4.2

CNT Distribution in a CNT Bundle of Width 32 nm and Thickness 60.8 nm (32-nm Technology Node)

CNT	Mean Dia. ↓ (nm)	Dia. → d (nm)							Total #CNTs
		0.9	1.0	1.1	1.2	1.3	1.4	1.5	
$d_\sigma = 0.1\ nm$									
Spacing = 0d	1.1	64	290	478	290	64	5	0	1191
	1.2	4	55	249	410	249	55	4	1026
	1.3	0	3	48	217	359	217	48	892
Spacing = 0.5d	1.1	17	184	402	184	17	0	0	804
	1.2	0	15	158	345	158	15	0	691
	1.3	0	0	13	138	301	138	13	603
Spacing = 1d	1.1	3	113	349	113	3	0	0	581
	1.2	0	3	97	299	97	3	0	499
	1.3	0	0	2	85	262	85	2	436
Spacing = 1.7d	1.1	0	52	292	52	0	0	0	396
	1.2	0	0	45	250	45	0	0	340
	1.3	0	0	0	39	219	39	0	297
$d_\sigma = 0.2\ nm$									
Spacing = 0d	1.1	153	223	253	223	153	82	34	1121
	1.2	71	133	193	219	193	133	71	1013
	1.3	28	67	127	184	209	184	127	926
Spacing = 0.5d	1.1	95	170	207	170	95	35	9	781
	1.2	30	80	145	176	145	80	30	686
	1.3	7	27	74	133	161	133	74	609
Spacing = 1d	1.1	57	133	177	133	57	14	1	572
	1.2	11	48	113	150	113	48	11	494
	1.3	1	10	43	101	134	101	43	433
Spacing = 1.7d	1.1	26	96	148	96	26	3	0	395
	1.2	2	22	82	126	82	22	2	338
	1.3	0	2	19	71	110	71	19	292

Note: Spacing between CNTs is expressed as percentage of CNT diameter.

Table 4.4 shows the values of resistance of CNT bundle for 32- and 16-nm technology nodes, respectively. As the spacing increases (i.e., as the porosity increases), the effective number of conducting channels decreases in a bundle and hence the resistance increases. The sigma variation in diameter has very little effect on bundle resistance.

Table 4.5 shows the kinetic inductance of CNT bundles for 32-nm technology node. With spacing, the kinetic inductance increases due to the reduction in effective number of conducting channels. The sigma

TABLE 4.3

CNT Distribution in a CNT Bundle of Width 16 nm and Thickness 32 nm (16-nm Technology Node)

CNT	Mean Dia. ↓ (nm)	Dia. → d (nm)							Total #CNTs
		0.9	1.0	1.1	1.2	1.3	1.4	1.5	
$d_\sigma = 0.1$ *nm*									
Spacing = 0d	1.1	17	76	126	76	17	1	0	313
	1.2	1	14	66	109	66	14	1	271
	1.3	0	1	12	58	95	58	12	236
Spacing = 0.5d	1.1	4	49	107	49	4	0	0	213
	1.2	0	4	42	91	42	4	0	183
	1.3	0	0	3	36	80	36	3	158
Spacing = 1d	1.1	1	30	92	30	1	0	0	154
	1.2	0	0	26	80	26	0	0	132
	1.3	0	0	0	22	69	22	0	113
Spacing = 1.7d	1.1	0	14	77	14	0	0	0	105
	1.2	0	0	12	66	12	0	0	90
	1.3	0	0	0	10	57	10	0	77
$d_\sigma = 0.2$ *nm*									
Spacing = 0d	1.1	40	59	67	59	40	21	9	295
	1.2	19	35	51	58	51	35	19	268
	1.3	7	18	33	49	55	49	33	244
Spacing = 0.5d	1.1	25	45	55	45	25	9	2	206
	1.2	8	21	38	47	38	21	8	181
	1.3	1	7	20	35	43	35	20	161
Spacing = 1d	1.1	15	36	48	36	15	3	0	153
	1.2	3	12	30	40	30	12	3	130
	1.3	0	2	11	27	36	27	11	114
Spacing = 1.7d	1.1	7	26	39	26	7	0	0	105
	1.2	0	6	22	33	22	6	0	89
	1.3	0	0	5	19	29	19	5	77

Note: Spacing between CNTs is expressed as percentage of CNT diameter.

variation in diameter has a very little effect on the kinetic inductance of a bundle.

Table 4.6 shows the magnetic inductance, which is almost constant for different CNT diameters and spacings. This is due to the mutual inductance between the CNTs in a bundle. In our analysis, we assumed that metallic CNTs are regularly spaced, which may not be true in real scenario. However, as magnetic inductance is almost the same for different CNT configurations in a bundle, it can be safely assumed that statistical

TABLE 4.4

Resistance (Ω) of CNT Bundle with Different CNT Diameter and Spacing Distribution, Length = 10 µm, $\lambda = 1$ µm

$d_\sigma \rightarrow$	0.1 nm			0.2 nm		
Spacing (%d) ↓	d_μ (nm)			d_μ (nm)		
	1.1	1.2	1.3	1.1	1.2	1.3
32-nm Technology						
0d	59.80	69.42	79.84	63.53	70.31	76.91
0.5d	88.58	103.07	118.11	91.19	103.82	116.95
1d	122.58	142.73	163.35	124.51	144.17	164.48
1.7d	179.85	209.47	239.80	180.31	210.71	243.91
16-nm Technology						
0d	227.54	262.81	301.78	241.43	265.75	291.89
0.5d	334.37	380.18	450.76	345.73	393.48	442.36
1d	462.47	539.55	630.27	465.49	547.85	624.74
1.7d	678.29	791.34	924.94	678.29	800.23	924.99

TABLE 4.5

Kinetic Inductance (pH) of CNT Bundle with Different CNT Diameter and Spacing Distribution, Length = 10 µm, $\lambda = 1$ µm

$d_\sigma \rightarrow$	0.1 nm			0.2 nm		
Spacing (%d) ↓	d_μ (nm)			d_μ (nm)		
	1.1	1.2	1.3	1.1	1.2	1.3
32-nm Technology						
0d	33.96	39.42	45.34	36.08	39.92	43.67
0.5d	50.30	58.53	67.07	51.78	58.95	66.41
1d	69.61	81.05	92.76	70.70	81.87	93.40
1.7d	102.13	118.95	136.17	102.38	119.65	138.50
16-nm Technology						
0d	129.21	149.23	171.36	137.09	150.90	165.75
0.5d	189.87	220.99	255.96	196.32	223.44	251.19
1d	262.61	306.38	357.89	264.33	311.09	354.75
1.7d	385.16	449.35	525.22	385.16	454.40	525.22

variation in positioning metallic CNTs will not have any significant effect on the magnetic inductance of the bundle. It is also found that the kinetic inductance of bundles is significantly higher than the magnetic inductance of the CNT bundle.

Tables 4.5 and 4.6 show the kinetic and magnetic inductances of CNT bundle for 16-nm technology nodes, respectively. In this case also, the

TABLE 4.6

Magnetic Inductance (pH) of CNT Bundle with Different CNT Diameter and Spacing Distribution, Length = 10 μm, λ = 1 μm

$d_\sigma \rightarrow$ Spacing (%d) ↓	0.1 nm d_μ (nm)			0.2 nm d_μ (nm)		
	1.1	1.2	1.3	1.1	1.2	1.3
32-nm Technology						
0d	11.53	11.57	11.63	11.61	11.57	11.53
0.5d	11.31	11.27	11.37	11.31	11.27	11.37
1d	11.16	11.14	11.13	11.16	11.14	11.13
1.7d	11.04	11.04	11.07	11.04	11.04	11.07
16-nm Technology						
0d	12.91	12.95	13.02	12.91	12.95	12.83
0.5d	12.63	12.70	12.80	12.63	12.70	12.80
1d	12.60	12.45	12.58	12.60	12.45	12.58
1.7d	12.35	12.51	12.36	12.35	12.51	12.36

magnetic inductance is almost invariant for different CNT diameters and spacings. It has also been found that the kinetic inductance of 16-nm technology nodes is almost four times more than that of 32-nm technology nodes.

This is due to the fact that the cross-sectional area for 16-nm technology nodes is almost four times smaller than that of 32-nm technology nodes, and hence, the former can accommodate fewer CNTs and fewer conducting channels.

Table 4.7 shows the quantum capacitance of CNT bundles for 32- and 16-nm technology nodes, respectively. With spacing, quantum capacitance decreases due to fewer effective conducting channels in a bundle. The sigma variation in diameter has little effect on the quantum capacitance of CNT bundles.

The variation in RLC values of CNT bundles of length 10 μm with different CNT diameters and spacings are shown in Figures 4.4 through 4.9.

It has been observed that as the CNT diameter is increased in a bundle of fixed width and thickness, the number of channels reduces, and hence, bundle resistance and inductance increase, and capacitance decreases.

For densely packed SWCNT bundles, electrostatic capacitance is the same as that of copper wire, as reported by Naeemi and Meindl [11]. For a porous bundle consisting of only four SWCNTs at the corners, capacitance is only 20% lesser than that of a densely packed bundle. Thus, it can be assumed that the electrostatic capacitance of SWCNT bundles is equal to that of copper wires [11].

TABLE 4.7

Quantum Capacitance (pF) of CNT Bundle with Different CNT Diameter
and Spacing Distribution, Length = 10 μm, λ = 1 μm

$d_\sigma \rightarrow$	0.1 nm			0.2 nm		
Spacing (%d) ↓	d_μ (nm)			d_μ (nm)		
	1.1	1.2	1.3	1.1	1.2	1.3
32-nm Technology						
0d	4.60	3.96	3.45	4.33	3.91	3.58
0.5d	3.11	2.67	2.33	3.02	2.65	2.35
1d	2.24	1.93	1.68	2.21	1.91	1.67
1.7d	1.53	1.31	1.15	1.53	1.31	1.13
16-nm Technology						
0d	1.21	1.05	0.91	1.14	1.04	0.94
0.5d	0.82	0.71	0.61	0.80	0.70	0.62
1d	0.60	0.51	0.44	0.59	0.50	0.44
1.7d	0.41	0.35	0.30	0.41	0.34	0.30

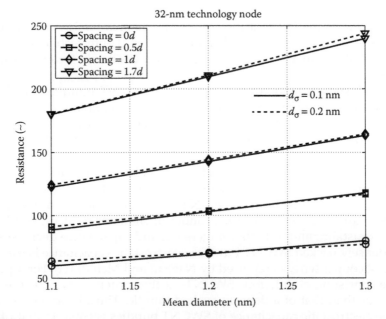

FIGURE 4.4
Resistance of CNT bundle with different CNT diameter and spacing for 32-nm technology node.

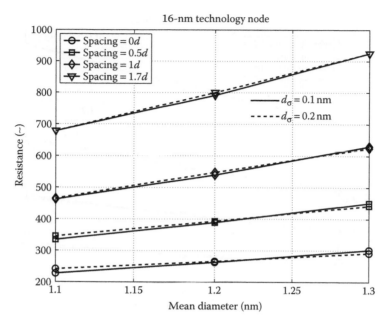

FIGURE 4.5
Resistance of CNT bundle with different CNT diameter and spacing for 16-nm technology node.

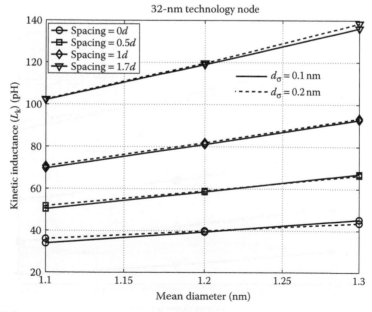

FIGURE 4.6
Kinetic inductance of CNT bundle with different CNT diameter and spacing for 32-nm technology node.

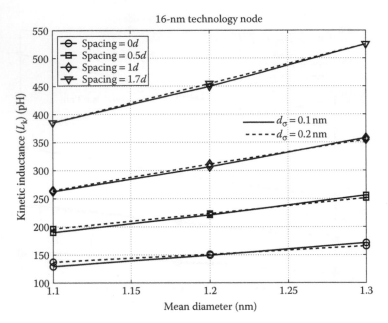

FIGURE 4.7
Kinetic inductance of CNT bundle with different CNT diameter and spacing for 16-nm technology node.

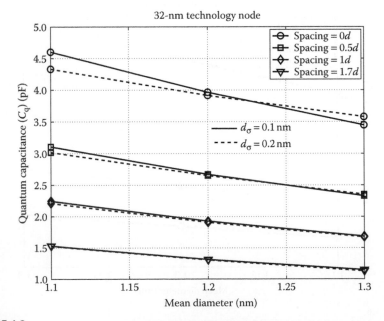

FIGURE 4.8
Quantum capacitance of CNT bundle with different CNT diameter and spacing for 32-nm technology node.

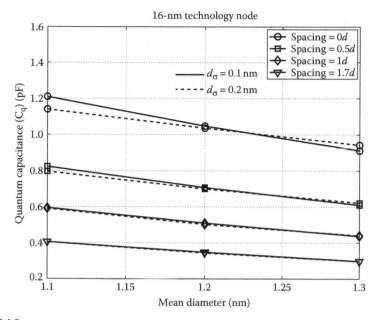

FIGURE 4.9
Quantum capacitance of CNT bundle with different CNT diameter and spacing for 16-nm technology node.

4.2.2 Modeling Temperature Variation

The total resistance of a CNT is given by

$$R_{CNT} = R_C + R_Q + R_O \tag{4.3}$$

where R_O is the temperature-dependent ohmic resistance as given in [20,21]:

$$R_O(T) = \frac{hl_{CNT}}{4e^2\lambda_{eff}(T)} \tag{4.4}$$

where l_{CNT} is the length and $\lambda_{eff}(T)$ is the effective electron mean free path [20–22], which is determined by the Matthiessen's rule as follows:

$$\frac{1}{\lambda_{eff}} = \frac{1}{\xi_{AC}} + \frac{1}{\xi_{OP,ems}} + \frac{1}{\xi_{OP,abs}} \tag{4.5}$$

Here, ξ_{AC} is the acoustic (AC) scattering length and $\xi_{OP,ems}$ and $\xi_{OP,abs}$ are the spontaneous optical (OP) emission lengths representing inelastic electron scattering caused by optical phonon emission and optical phonon absorption, respectively.

Temperature dependence is modeled [21] by expressing the scattering length as a function of temperature as

$$\xi_{AC} = \xi_{AC,300}\left(\frac{300}{T}\right) \tag{4.6}$$

where $\xi_{AC,300}$ is the acoustic phonon scattering length at 300 K (= 1600 nm). According to [34], $\xi_{AC} = 4.8 \times 10^{-4}/T$, at any temperature T. In [22,24], a diameter-dependent acoustic phonon scattering length is given as

$$\xi_{AC} = 400 \times 10^3 \times \left(\frac{d}{T}\right) \tag{4.7}$$

For $d = 1$ nm, $\xi_{AC} = 4.0 \times 10^{-4}/T$ according to Equation 4.7.

The optical phonon absorption length is given by [21,22,24]:

$$\xi_{OP,abs} = \xi_{OP}\frac{N_{OP}(300)+1}{N_{OP}(T)} \tag{4.8}$$

where $\xi_{OP} \sim 15$ nm is the spontaneous optical phonon emission length at 300 K [21]. However, a diameter-dependent expression is $\xi_{OP} = 56d$, whereas measurements indicate smaller coefficients ($\sim$15–20) [22]. Therefore, we assume $\xi_{OP} = 15$ nm in our analysis considering nanotube diameter to be 1 nm.

$$N_{OP} = \left[\frac{1}{e^{\hbar\omega_{OP}/k_BT}-1}\right] \tag{4.9}$$

The parameter N_{OP} denotes the optical phonon occupation

where:

$\hbar\omega_{OP} \sim$0.16–0.2 eV is the OP emission threshold energy
k_B is the Boltzmann constant

The optical phonon emission is contributed by two processes. One is the applied electric field and the other is the absorption of an optical phonon. For the electric field-induced phonon emission, the OP emission scattering length [22] is given by

$$\xi_{OP,ems}^{fld} = \frac{\hbar\omega_{OP} - k_BT}{eV / l_{CNT}} + \frac{N_{OP}(300)+1}{N_{OP}(T)+1}\xi_{OP} \tag{4.10}$$

where V is the applied bias. The first term represents OP emission length due to the applied electric field ($E = V/l_{CNT}$) and the second term represents the temperature-dependent OP emission length. Similarly, for the phonon emission due to the optical phonon absorption, the OP emission scattering length is given by

$$\xi_{OP,ems}^{abs} = \xi_{OP,abs} + \frac{N_{OP}(300)+1}{N_{OP}(T)+1}\xi_{OP} \tag{4.11}$$

Therefore, the OP emission mean free path due to the combined effect of applied electric field and OP absorption [20] is given by

$$\frac{1}{\xi_{OP,ems}} = \frac{1}{\xi_{OP,ems}^{fld}} + \frac{1}{\xi_{OP,ems}^{abs}} \tag{4.12}$$

Substituting Equations 4.10 and 4.11 in Equation 4.12, we obtain

$$\xi_{OP,ems} = \frac{\left[\left(\hbar\omega_{OP}/eV_{DD}\right)l + \left\{[1+N_{OP}(300)]/[1+N_{OP}(T)]\right\}\xi_{OP}(300)\right]\left[1+2N_{OP}(T)\right]}{1+3N_{OP}(T)+\left\{\hbar\omega_{OP}N_{OP}(T)[1+N_{OP}(T)]/eV_{DD}\xi_{OP,300}[1+N_{OP}(300)]\right\}l} \tag{4.13}$$

Hence, the effective mean free path can be obtained by substituting Equations 4.6, 4.8, and 4.13 in Equation 4.5.

Figures 4.10 and 4.11 depict the temperature variation of resistance of CNT bundle-based interconnects for 32- and 16-nm technology nodes. The temperature-dependent resistance of CNT bundle is shown in Table 4.8. With increases in temperature, the bundle resistance increases nonlinearly, unlike copper wires. Therefore, CNT-based wires are more temperature sensitive compared to copper wires.

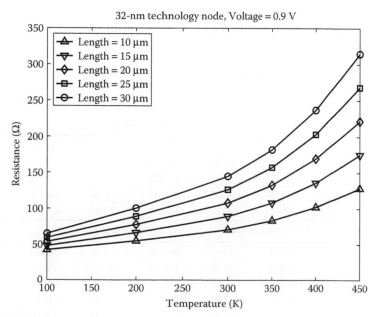

FIGURE 4.10
Resistance of CNT bundle versus temperature for 32-nm technology node.

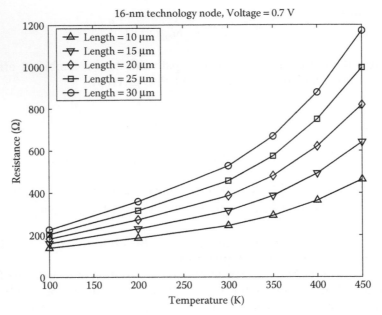

FIGURE 4.11
Resistance of CNT bundle versus temperature for 16-nm technology node.

TABLE 4.8

Resistance (Ω) of CNT Bundle Interconnect of Different Length at Different Temperature

Temperature (K) →	100	200	300	350	400	450
Length (μm) ↓			**32-nm Technology**			
10	42.15	54.73	70.48	83.28	102.10	128.52
15	47.87	66.15	89.13	107.98	135.84	175.09
20	53.56	77.55	107.75	132.65	109.55	221.63
25	59.24	88.92	126.36	157.30	203.24	268.14
30	64.92	100.30	144.95	181.94	236.92	314.65
Length (μm) ↓			**16-nm Technology**			
10	139.00	185.90	244.80	292.90	363.80	463.70
15	160.70	229.30	315.60	386.70	492.10	640.80
20	182.30	272.60	386.40	480.50	620.30	817.80
25	203.90	315.80	457.20	574.30	748.50	994.80
30	225.50	359.10	527.90	668.10	876.60	1171.70

4.2.3 Modeling Voltage Variation

The resistance of SWCNT depends on applied voltage. For nanotubes operating under a low bias voltage ($V_b < 0.1$ V), the resistance is independent of applied bias voltage and is expressed as

$$R_{\text{low}} = R_Q + R_O + R_C, \quad \text{if } l_{\text{CNT}} > \lambda_{\text{ef}} \tag{4.14}$$

Under a large bias voltage, resistance depends on the applied bias voltage as

$$R_{\text{high}} = R_{\text{low}} + \frac{V_{\text{BIAS}}}{I_{\text{MAX}}} \tag{4.15}$$

where I_{MAX} is the maximum current that can flow through a CNT, which is approximately 25 µA [3].

In our work, the applied bias is high ($V_{\text{BIAS}} > 0.1$ V), and hence, a combination of Equations 4.3, 4.4, and 4.13 is used to model the voltage dependency of the CNT resistance.

Figures 4.12 and 4.13 show the voltage dependency of resistance of CNT bundle interconnects with interconnect dimensions corresponding to 32- and 16-nm technology nodes. The voltage-dependent resistance of CNT bundles is shown in Table 4.9. With the increase in operating voltage, the

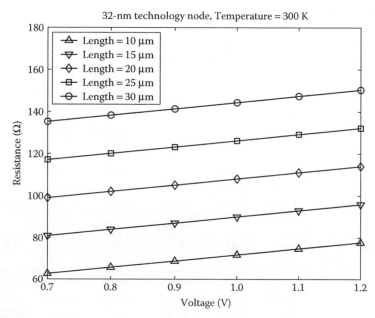

FIGURE 4.12
Resistance of CNT bundle versus voltage for 32-nm technology node.

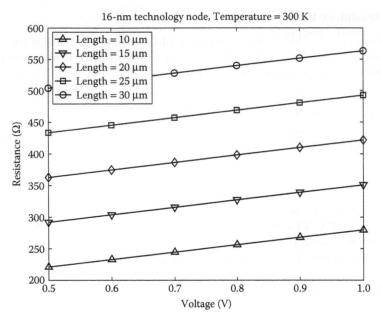

FIGURE 4.13
Resistance of CNT bundle versus voltage for 16-nm technology node.

TABLE 4.9

Resistance (Ω) of CNT Bundle Interconnect of Different Length at
Different Operating Voltage, Temperature: 300 K

Voltage (V) → Length (μm) ↓	32-nm Technology					
	0.7	0.8	0.9	1.0	1.1	1.2
10	64.32	67.40	70.48	73.55	76.62	79.68
15	82.95	86.04	89.13	92.22	95.31	98.39
20	101.55	104.65	107.75	110.85	113.94	117.03
25	120.15	123.25	126.36	129.46	132.56	135.65
30	138.74	141.84	144.95	148.06	151.16	154.26
Voltage (V) → Length (μm) ↓	16-nm Technology					
	0.5	0.6	0.7	0.8	0.9	1.0
10	221.24	233.01	244.76	256.48	268.19	279.88
15	292.03	303.83	315.63	327.40	339.17	350.92
20	362.78	374.60	386.41	398.22	410.01	421.79
25	433.51	445.34	457.17	468.99	480.80	492.60
30	504.23	516.07	527.91	539.73	551.56	563.37

bundle resistance increases linearly. Therefore, CNT-based wires are voltage sensitive, unlike copper wires.

4.2.4 Modeling and Analysis Based on Random Distribution of SWCNTs

In a bundle of SWCNT, there are 33% metallic CNTs and rest are semiconducting. For interconnect applications, metallic tubes alone are important as they are responsible for current conduction. The locations of the metallic tubes vary from process to process. In [33], the authors investigated the impact of the positional variation of metallic CNTs in a bundle on the electrical circuit parameters and hence on the timing delay.

In [33], the authors considered three different random unique distributions of CNTs in a SWCNT bundle. They considered two cases: (1) The diameter of all the CNTs in the SWCNT bundle is considered to be fixed (1 nm) as illustrated in Figure 4.14; (2) the varying diameter of the CNTs is considered where the diameter variation follows the Gaussian distribution with mean diameter of 1.1 nm (see Figure 4.15).

The analysis of SWCNT bundles is carried out for three different random distributions of CNTs within a bundle considering a fixed diameter of CNTs as well as for CNTs having variable diameter. It has been observed that the resistance of SWCNT bundles remains unchanged for any statistical distribution of CNTs within a bundle. The same observation holds true for capacitance and kinetic inductance of CNT bundle. The only parameter that changes its value with the change in distribution of CNTs is magnetic

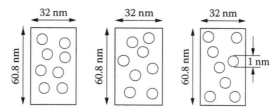

FIGURE 4.14
Random distribution of SWCNTs in a bundle with all SWCNTs having a fixed diameter.

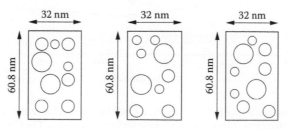

FIGURE 4.15
Random distribution of SWCNTs of variable diameter in a bundle.

inductance, which we can observe from Equation 3.36 in Chapter 3, where the mutual inductance depends on the variable ρ, that is, the center-to-center spacing between two nanotubes, which will clearly change if we consider various CNT distributions.

4.2.5 Results and Discussion

In order to compare the performance of interconnect with densely packed CNT bundles and copper wire, we calculated the delay through interconnects of different lengths for different technology nodes. Figure 4.16 shows the interconnect delay ratio between CNT- and copper-based interconnects.

It is observed from Figure 4.16 that the delay through the interconnect with densely packed CNT bundles is higher at shorter length. This is due to the large quantum resistance of the CNT, which results in higher per-unit-length resistance at shorter lengths.

It is also observed from Figure 4.16 that as the interconnect dimension is scaled down, the delay improvement of CNT bundles over copper wire becomes significant and minimum length for which τp(CNT):τp(Cu) < 1.0 is reduced. Therefore, the CNT bundle has better delay performance over copper wire for intermediate (>10 μm) and global (>100 μm) interconnects, as predicated by earlier works [11,12]. The performance is improved with the scaling of interconnect dimensions.

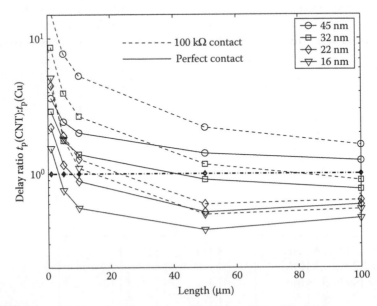

FIGURE 4.16
Ratio of interconnect delay with densely packed CNT bundle to that with copper interconnect as a function of interconnect length for different technology nodes. The CNT bundle assumes densely packed CNTs with equal diameter of 1 nm. Bold dotted line indicates delay ratio = 1.0, signifies delay through copper and CNT is same.

TABLE 4.10

Interconnect Delay (ps) of CNT Bundle Interconnect for Different Process Variation, Length = 10 µm, V_{DD} = 0.9 V (32 nm) and 0.7 V (16 nm), T = 300 K

d_σ = 0.1 nm Spacing (%d) ↓	32-nm Technology d_μ (nm)			16-nm Technology d_μ (nm)		
	1.1	1.2	1.3	1.1	1.2	1.3
0d	0.026	0.031	0.037	0.052	0.052	0.059
0.5d	0.043	0.055	0.067	0.072	0.105	0.137
1d	0.069	0.078	0.078	0.142	0.152	0.147
1.7d	0.077	0.086	0.117	0.147	0.165	0.210

Table 4.10 shows the interconnect delay for different CNT diameters and spacings in a bundle corresponding to 32- and 16-nm technology nodes. It is found that with increase in spacing, delay increases significantly. For a sparsely packed bundle with P_m = 1/3, the delay increases almost three times as compared to the densely packed bundle. As the mean diameter of CNTs in a bundle increases, the delay increases by ~50% for a variation of 1.1–1.3 nm. However, for a sparsely packed bundle with P_m = 1/2, diameter variation has almost no effect on the delay for both 32- and 16-nm technology nodes. It is found that with the increase in mean diameter of CNTs in a sparsely packed bundle with P_m = 1/2, the increase in resistance and inductance and decrease in capacitance make the series resonant frequency ($\omega_n = 1/\sqrt{LC}$) and damping factor ($\xi = 0.5\sqrt{C/L}$) [31] almost equal, which in turn makes the delay through CNT interconnect almost equal. Hence, a sparsely packed bundle with 50% metallic CNTs (P_m = 1/2) along with diameter variation is better from a process-technology point of view as interconnect delay is process invariant.

The delay through 10-µm-long CNT interconnects with dimensions corresponding to 32- and 16-nm technology nodes is shown in Figures 4.17 and 4.18 for different CNT diameters and spacings.

Table 4.11 shows the interconnect delay under different values of PTV. The densely packed bundle corresponds to d_μ = 1.1 nm and $d\sigma$ = 0.1 nm with P_m = 1. The moderately packed bundle corresponds to d_μ = 1.2 nm and $d\sigma$ = 0.1 nm with P_m = 2/3. The sparsely packed bundle corresponds to d_μ = 1.2 nm and $d\sigma$ = 0.1 nm with P_m = 1/3. The delay is minimum for densely packed bundles operating under low temperature and maximum for sparsely packed bundle operating under high temperature. The delay variation with operating voltage is within approximately ±20% from the average value for both technology nodes.

With temperature variation, the delay variation is ±60% for 32-nm technology node and ±50% for 16-nm technology node from the average value. With

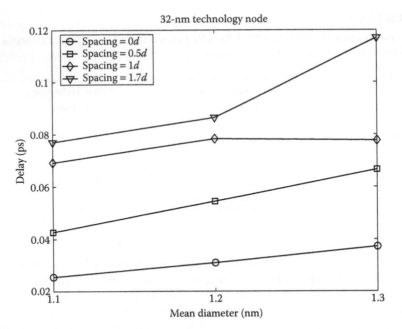

FIGURE 4.17
Propagation delay through CNT bundle with different CNT diameter and spacing. The delay values are calculated at nominal V_{DD} and room temperature 300 K. The interconnect length is 10 µm and $w = 32$ nm and $t = 60.8$ nm for 32-nm technology node.

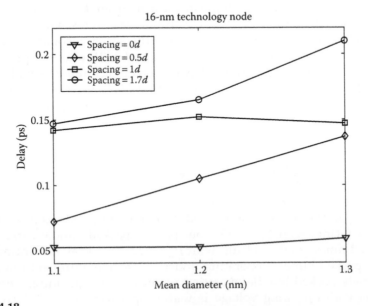

FIGURE 4.18
Propagation delay through CNT bundle with different CNT diameter and spacing. The delay values are calculated at nominal V_{DD} and room temperature 300 K. The interconnect length is 10 µm and $w = 16$ nm and $t = 32$ nm for 16-nm technology node.

TABLE 4.11

Interconnect Delay (fs) of CNT Bundle Interconnect for Different PTV Condition, Length = 10 μm.

Temperature (K) ↓	Densely Packed			Moderately Packed			Sparsely Packed		
	32-nm Technology								
Voltage (V) →	0.8	0.9	1.0	0.8	0.9	1.0	0.8	0.9	1.0
200	34	26	30	53	42	47	71	127	139
300	28	32	43	61	64	66	137	144	172
400	50	53	58	85	94	102	199	221	240
	16-nm Technology								
Voltage (V) →	0.6	0.7	0.8	0.6	0.7	0.8	0.6	0.7	0.8
200	32	60	47	80	92	89	136	98	116
300	57	57	71	105	112	104	243	232	288
400	87	94	103	150	163	179	357	391	422

a different metallic fraction (P_m), the delay variation is significant: −65% to 120% for 32-nm technology node and −60% to 105% for 16-nm technology node from the average value.

Figures 4.19 and 4.20 show the delay variation under different PTV conditions for 32- and 16-nm technology nodes. With voltage variation, the delay

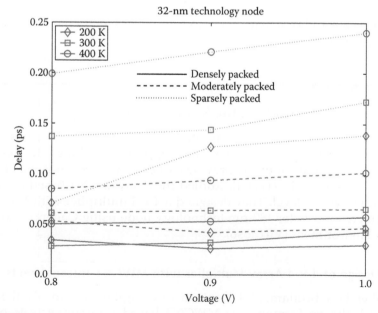

FIGURE 4.19

(See color insert.) Propagation delay through CNT bundle at different PTV. The interconnect length is 10 μm and technology node is 32 nm.

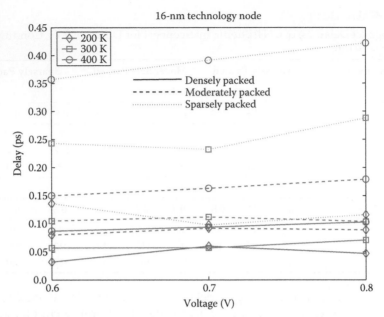

FIGURE 4.20
(See color insert.) Propagation delay through CNT bundle at different PTV. The interconnect length is 10 μm and technology node is 16 nm.

variation is insignificant, whereas with process and temperature variations, the delay variation is significant.

4.3 Performance Study at the System Level

It has already been found that multiwalled CNT (MWCNT) has superior performance over the conventional copper net for long interconnects [35–39]. However, most of the previous studies focused on isolated CNT interconnects. In [40], Das et al. studied the delay of CNT-based interconnects at the system level. They designed a 4 × 4 multiplier with MWCNT-based interconnects for 180-nm technology node using Cadence® Analog Design Environment (Virtuoso).

4.3.1 Design of 4 × 4 Array Multiplier with MWCNT Interconnects

Multiplier is a fundamental block in any digital system. At the system level, the performance of MWCNT-based interconnects is investigated, by designing a 4-bit unsigned array multiplier using MWCNT interconnects.

4.3.2 Basic Design Block

A schematic of a 4×4 unsigned array multiplier and its basic building block is shown in Figures 4.21 and 4.22. The 4×4 multiplier has two binary inputs: multiplicand (A) and multiplier (B). Here inputs A and B each have 4-bits. Let us assume them as a_3 down to a_0 and b_3 down to b_0. The output P consists of 8-bits expressed as P_7 down to P_0. The product can be written as

$$P = \sum_{i=0}^{3} a_i 2^i \times \sum_{j=0}^{3} b_j 2^j \tag{4.16}$$

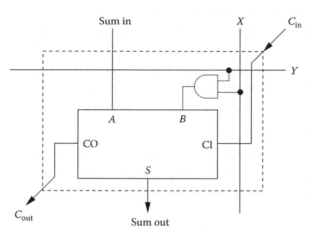

FIGURE 4.21
Basic building block of a 4×4 unsigned array multiplier.

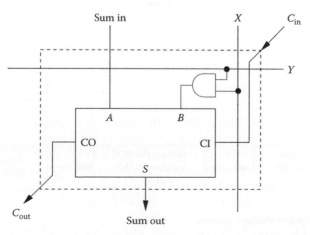

FIGURE 4.22
Schematic of a 4×4 unsigned array multiplier.

4.3.3 Design Methodology

The design methodology is explained with the help of the flow chart shown in Figure 4.23.

Step 1. Design the schematic of 4 × 4 array multiplier in Cadence Analog Design Environment using Virtuoso or using similar schematic editor.

Step 2. Next, design the layout of the 4 × 4 multiplier using Layout-XL based on gpdk 180 nm or using the available technology.

Step 3. Collect the interconnect parameters: length, width, and layer number of each net from the layout.

Step 4. Model each net by MWCNT bundle and insert its RLC values in the net list.

Step 5. Simulate the design using Cadence Spectre Circuit Simulator or a similar circuit simulation tool to obtain the results.

4.3.4 Layout Design

The layout of the unsigned multiplier is shown in Figure 4.24, which is drawn based on the gpdk-180 nm design rule in Cadence Analog Design Environment by using Layout-XL.

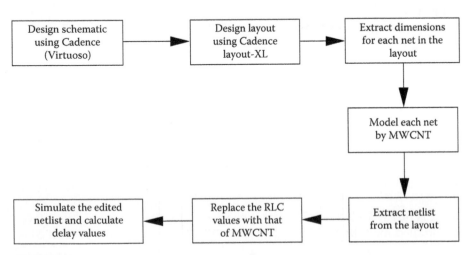

FIGURE 4.23
Flow chart of multiplier design process.

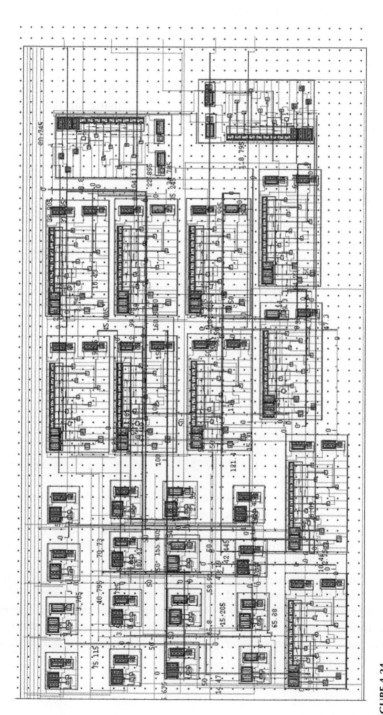

FIGURE 4.24
Layout of 4 × 4 unsigned array multiplier.

4.3.5 Extraction of RLC Parameters

The RLC parameters of the interconnects for both copper and MWCNT are extracted as follows. Information on length of the interconnects is obtained from the layout. Using the layer information, the interconnect width and thickness are decided for the 180-nm process technology. The RLC values of the copper-based interconnects are then obtained using the Equations 3.24 through 3.51 in Chapter 3. Next each of the interconnects is replaced by the MWCNT double bundle as explained in Sections 4.3.3 and 4.3.4. The RLC values of the MWCNT-based interconnects are then obtained using the Equations 3.21 through 3.23 in Chapter 3.

4.3.6 Simulation Results

The simulation setup is shown in Figure 4.25, where the 4×4 multiplier is simulated with 8 input pulses. The input and output waveforms of the test circuit are shown in Figure 4.26. In this analysis, the delay through each net is calculated as well as the overall system delay.

The layout of the unsigned multiplier is shown in Figure 4.24, which is drawn based on the gpdk-180 nm design rule in Cadence Analog Design Environment by using Layout-XL.

Table 4.12 shows the details of each net obtained from the drawn layout of the 4×4 multiplier. Using the dimensions of each net in the layout, the nets are modeled using MWCNTs employing the methodology described earlier. At the system level, the performance of the MWCNTs is studied based on the delay ratio of MWCNT-based interconnects and the conventional copper-based interconnects.

Table 4.13 shows the actual net delays for copper- and MWCNT-based interconnects. It is shown that for all the nets, the delay ratio ($= \tau_{MWCNT}{:}\tau_{Cu}$) is less than 1.0. This indicates that MWCNT-based nets have less delay compared to copper-based nets.

Table 4.14 shows how the overall system delay changes when copper-based nets are replaced by MWCNT-based nets. In this design, the critical path is $a_0 \rightarrow P_7$, which has the maximum delay. The critical path delay is reduced from 5.403 to 5.22 ns (3.4% reduction) when the copper nets are replaced by MWCNT in the critical path.

From this study, it is found that the MWCNT-based interconnect system has lesser delay than conventional copper wire in system level at 180-nm technology node. The methodology of replacing copper nets by CNT nets explained in this chapter provides a simple and viable approach in modeling and analyzing the CNT-based interconnect systems using the conventional chip design flow and tools.

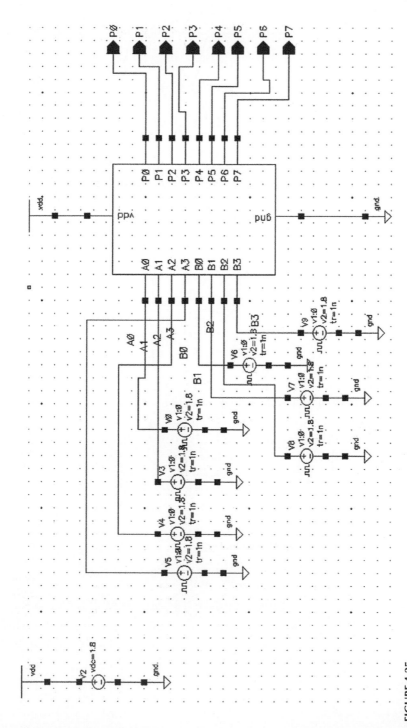

FIGURE 4.25
Simulation setup for the 4 × 4 multiplier.

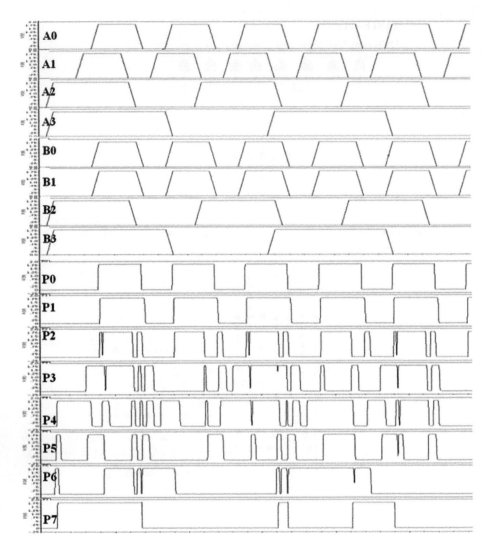

FIGURE 4.26
Input and output waveforms of the 4 × 4 multiplier.

TABLE 4.12

Different Net Names, Metalization Layers, and Their Lengths

Net Name	Length (µm)	Metal Layers
Net 34	153.96	Layer 1
Net 39	293.065	Layer 1, Layer 2
Net 45	330.975	Layer 1, Layer 2
Net 53	369.91	Layer 1, Layer 2
Net 68	283.815	Layer 1, Layer 2
Net 76	132.88	Layer 1
Net 93	172.425	Layer 1
Net 162	213.31	Layer 1, Layer 2
Net 130	129.63	Layer 1
Net 111	132.175	Layer 1, Layer 2
Net 138	167.68	Layer 1
Net 157	204.92	Layer 1, Layer 2
Net 177	194.315	Layer 1, Layer 2
Net 190	229.165	Layer 1, Layer 2
Net 201	226.155	Layer 1, Layer 2
Net 219	64.56	Layer 1
Net 171	66.165	Layer 1
Net 100	136.89	Layer 1, Layer 2
Net 221	134.56	Layer 1
Net 218	201.35	Layer 1, Layer 2
Net 220	196.34	Layer 1, Layer 2
Net 123	102.775	Layer 1
Net 155	96.905	Layer 1
Net 194	98.71	Layer 1
Net 89	60.945	Layer 1
Net 128	62.465	Layer 1
Net 165	59.604	Layer 1
Net 51	281.995	Layer 1
Net 98	281.36	Layer 1
Net 127	310.025	Layer 1
Net 63	130.55	Layer 1
Net 91	135.08	Layer 1

TABLE 4.13

Delay Values of Interconnects Drawn in the Layout of the Multiplier

Length (μm)	Delay (τ) in ps		Delay Ratio τ_{MWCNT}/τ_{Cu}
	Copper	MWCNT	
59.604	9.10e-01	5.67e-01	0.62
60.945	9.39e-01	5.70e-01	0.61
62.465	9.66e-01	5.74e-01	0.59
64.56	1.01e-00	5.96e-01	0.59
66 165	1.04e-00	6 14e-01	0.59
96.905	1.60e-00	8.62e-01	0.54
98.71	1.64e-00	8.66e-01	0.53
102.755	1.72e-00	8.74e-01	0.51
129.63	2.28e-00	1.12e-00	0.49
130.55	2.29e-00	1 10e-00	0.48
132.175	2.34e-00	1.11e-00	0.47
132.88	2.36e-00	1.11e-00	0.47
134.56	2.40e-00	1.12e-00	0.47
135.08	2.44e-00	1 13e-00	0.46
136.89	2.54e-00	1.17e-00	0.46
153.96	2.85e-00	1.30e-00	0.46
167.68	3.19e-00	1.44e-00	0.45
172.425	3.26e-00	1.47e-00	0.45
194.315	3.92e-00	1.75e-00	0.45
196.34	3.98e-00	1.76e-00	0.44
201.35	4.15e-00	1.83e-00	0.44
204.92	4.23e-00	1.86e-00	0.44
213.31	4.29e-00	1.88e-00	0.44
226.155	4.45e-00	1.94e-00	0.44
229.165	4.46e-00	1.94e-00	0.43
281.36	4.93e-00	2.12e-00	0.43
281.995	4.94e-00	2.12e-00	0.43
283.815	4.09c-00	2.14e-00	0.43
293.065	5.22e-00	2.23e-00	0.43
310.025	5.36e-00	2.28e-00	0.43
330.975	6.14e-00	2.61e-00	0.43
360.91	6.75e-00	2.82e-00	0.42

TABLE 4.14

Overall System Delay

Path	System Delay (μns) with	
	Copper Interconnects	**MWCNT Interconnects**
$a_0 \rightarrow P_7$	5.403	5.22
$a_0 \rightarrow P_3$	1.511	1.434
$a_1 \rightarrow P_4$	3.315	3.23
$a_1 \rightarrow P_7$	3.403	3.228
$u_2 \rightarrow P_3$	4.489	4.566
$a_3 \rightarrow P_2$	4.489	4.566
$b_0 \rightarrow P_3$	1.511	1.434
$b_1 \rightarrow P_4$	3.315	3.23
$b_1 \rightarrow P_7$	3.403	3.228
$b_2 \rightarrow P_3$	4.489	4.566
$b_3 \rightarrow P_2$	4.489	4.566

4.4 Summary

In this chapter, we presented the performance study of CNT-based interconnects considering PTV variations. The performance of CNT-based interconnects is investigated at the system level by designing a 4 × 4 unsigned multiplier using MWCNT-based interconnects. The results show that CNT interconnects have superior performance over copper interconnects with respect to timing, RF response, and system-level perspective.

5

RF and Stability Analyses in CNT and GNR Interconnects

5.1 Introduction

The performance of carbon nanotube (CNT) based interconnects is better than that of traditional copper-based interconnects, as we have seen in Chapter 4. There are several other works that show superiority of GNR interconnects over copper interconnects (references 16 and 17 in Chapter 3; reference 3 in Chapter 7). It is very important to investigate the high-frequency response of the CNT- and GNR-based interconnects, in addition to the timing performance of the CNT- and GNR-based interconnects. At very high frequencies, traditional interconnects experience the skin effect. The resistance and inductance increase due to the skin effect at higher frequencies.

5.2 RF Performance Analysis

Besides their circuit performance in terms of delay, performance of CNTs at high frequencies makes them a promising candidate for future interconnects. In this chapter, we develop a frequency-dependent model of CNT-based interconnects. Using the frequency-dependent model, the equivalent circuit parameters are extracted and the radio frequency (RF) performance is studied in terms of the scattering (S) parameters using a simulation program with integrated circuit emphasis (SPICE) based methodology.

The performance of CNT and GNR interconnects was compared with that of copper wire, and it was found that CNTs have RF performance comparable with copper wires. The simulated results are also compared with the experimentally characterized RF performance of CNTs available in different literatures.

5.2.1 Frequency-Dependent Models of Resistance and Inductance for Copper Wire

It is known that the resistance and inductance of the traditional metal conductors have strong dependency on the operation frequency due to the skin effect and proximity effect. The resistance increases and the inductance decreases with the increase in frequency. At high frequencies, the electrons tend to flow along the surface of the conductors rather than through the entire cross-sectional area of the conductor. This leads to increase in the resistance of the wire due to the reduction in effective cross-sectional area of the conductor. The expression for the per-unit-length resistance as a function of frequency can be expressed as

$$R(f) = \frac{\rho}{2(w+t)\delta} \tag{5.1}$$

where:
ρ is the resistivity of the conductor material
w and t are the conductor width and thickness, respectively

The parameter δ is called skin depth, which is defined as the depth of penetration, at which the current density is attenuated by 1 neper (1/e) and is given by

$$\delta(f) = \sqrt{\frac{\rho}{\pi f \mu_0}} \tag{5.2}$$

where:
f is the operation frequency
$\mu_0 = 4\pi \times 10^{-7}$ H/m is the conductor permeability

The expression for the frequency-dependent inductance can be found in [1]. Figures 5.1 and 5.2 show the resistance and inductance of copper wire of length 10 μm for different technology nodes, as a function of operation frequency (extracted using FastHenry [2]).

Considering the skin effect, the analytical expression for the frequency-dependent inductance [3] is given by

$$L_s(f) = \frac{\mu_0 l}{2\pi} \left[\ln\left(\frac{2l}{w+t}\right) + \frac{1}{2} + \frac{0.2235(w+t)}{l} - \mu_r(0.25 - X) \right] \tag{5.3}$$

The skin depth term X is expressed by

$$X = \begin{cases} 0.4372x, & \text{if } x < 0.5 \\ 0.0578x + 0.1897, & \text{if } 0.5 \leq x \leq 1.0 \\ 0.25, & \text{if } x > 1.0 \end{cases} \tag{5.4}$$

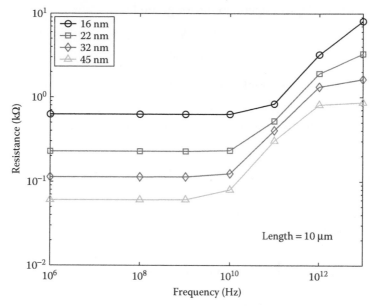

FIGURE 5.1
(See color insert.) Resistance of copper interconnect versus frequency for different technology nodes.

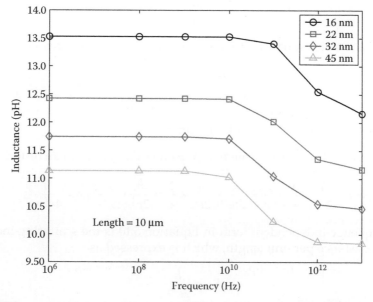

FIGURE 5.2
(See color insert.) Inductance of copper interconnect versus frequency for different technology nodes.

where x is a skin depth-dependent factor given by

$$x = \frac{\delta}{0.2235(w+t)} \tag{5.5}$$

5.2.2 Frequency-Dependent CNT Model

In order to find out the consequences of skin effect in CNTs, we consider the AC electrical conductivity according to the Drude model in [4]:

$$\sigma(\omega) = \frac{\sigma_0}{1 + j\omega\tau} \tag{5.6}$$

where:

τ is the momentum relaxation time
ω is the frequency in radian
σ_0 is the DC electrical conductivity

In terms of resistivity, we can also write Equation 5.6 as

$$\rho(\omega) = \frac{1 + j\omega\tau}{\sigma_0} = \frac{1}{\sigma_0} + j\omega\frac{\tau}{\sigma_0} \tag{5.7}$$

For a one-dimensional conductor having one conduction channel and considering only spin degeneracy [5], σ_0 can be written as

$$\sigma_0 = \frac{2e^2}{h} \times \lambda \tag{5.8}$$

where λ is the electron mean free path, which in CNT is expressed as

$$\lambda = v_F \times \tau_B \tag{5.9}$$

where:

v_F is the Fermi velocity
τ_B is the momentum backscattering time

In CNT, the backscattering time $\tau_B = 2\tau$. Hence, Equation 5.7 can be written as

$$\rho(\omega) = \frac{h}{2e^2} \times \frac{1}{\lambda} + j\omega \times \frac{\lambda}{2v_F} \times \frac{h}{2e^2} \times \frac{1}{\lambda} = \frac{h}{2e^2} \times \frac{1}{\lambda} + j\omega \times \frac{h}{4e^2v_F} \tag{5.10}$$

The frequency-independent term in Equation 5.10 is the scattering-induced ohmic resistance per unit length, which is expressed as

$$R_O = \frac{h}{2e^2\lambda} \tag{5.11}$$

and the second term in Equation 5.5 is modeled as inductive reactance ($j\omega L_K$), where L_K is the kinetic inductance per unit length, which is expressed as

$$L_K = \frac{h}{4e^2 v_F} \tag{5.12}$$

The electromagnetic wave propagating through the interconnect has the electric field component expressed as [6]

$$\frac{d^2 E_z}{dz^2} = \vartheta^2 E_z \tag{5.13}$$

where ϑ is the propagation constant given by

$$\vartheta = \sqrt{j\omega\mu\sigma} = \alpha + j\beta \tag{5.14}$$

We considered that for a conducting medium $\sigma \gg \omega\varepsilon$. Substituting the frequency-dependent conductivity from Equation 5.6 in Equation 5.14 and solving for α and β, we get

$$\alpha = \sqrt{\frac{\omega\mu\sigma_0}{2}} \times \sqrt{\frac{1}{1+(\omega\tau)^2} \left[\sqrt{1+(\omega\tau)^2} + \omega\tau \right]} \tag{5.15}$$

$$\beta = \sqrt{\frac{\omega\mu\sigma_0}{2}} \times \sqrt{\frac{1}{1+(\omega\tau)^2} \left[\sqrt{1+(\omega\tau)^2} - \omega\tau \right]} \tag{5.16}$$

The skin depth of the CNT bundle can be written as

$$\delta(\omega) = \frac{1}{\alpha} = \sqrt{\frac{2}{\omega\mu\sigma_0}} \times \sqrt{\left[1+(\omega\tau)^2 \right] \times \left[\sqrt{1+(\omega\tau)^2} - \omega\tau \right]} \tag{5.17}$$

For conductors τ is very small, therefore, $\omega\tau \ll 1$. So the expression of skin depth reduces to

$$\delta(\omega) = \sqrt{\frac{2}{\omega\mu\sigma_0}} \tag{5.18}$$

For CNTs, the momentum relaxation time is given by

$$\tau = \frac{\tau_B}{2} = \frac{\lambda}{2v_F} = \frac{500 \times D}{v_F} \tag{5.19}$$

Using Equations 5.17 and 5.19, one can calculate the skin depth of CNTs. Figure 5.3 shows the skin depth of copper- and CNT-based interconnects for frequencies ranging from 1 to 4000 GHz. It is observed that unlike copper wire, where skin depth monotonically decreases with increase in frequency, the skin depth of CNT interconnects tends to saturate after certain frequencies due to large momentum relaxation time (τ). The saturation frequency is again different

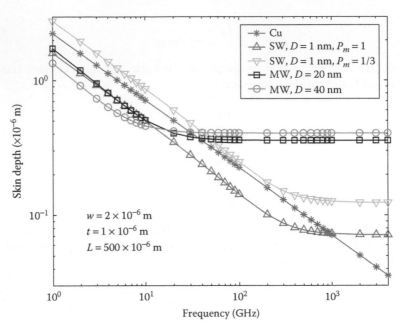

FIGURE 5.3
(**See color insert.**) Skin depth of copper and different CNT interconnect as a function of frequency. PM is the metallic fraction in CNT bundle.

for single-walled CNTs (SWCNTs) and multiwalled CNTs (MWCNTs). In case of MWCNTs, the saturation frequency is much lower compared to SWCNTs, which indicates that MWCNTs would be a better choice for high-frequency applications. Using the expression for skin depth in Equation 5.17, we can calculate the frequency-dependent resistance for CNT-based interconnects. Figures 5.4 and 5.5 show the resistance and inductance as a function of frequency.

5.2.3 Frequency-Dependent GNR Model

In [7], Sarkar et al. developed a similar as CNT interconnect frequency-dependent GNR model. The expression for skin depth in GNR interconnects is derived, which is same as Equation 5.17. The variation of skin depth with frequency is shown in Figure 5.6.

Considering the normal skin effect and Ohm's law are valid, surface resistance as a function of frequency could be expressed as:

$$Z_S = \sqrt{\frac{\omega\mu_0}{2\sigma}}\left[\sqrt{\sqrt{(\omega\tau)^2+1}-\omega\tau}+j\sqrt{\sqrt{(\omega\tau)^2+1}+\omega\tau}\right] \quad (5.20)$$

Using Equation 5.20, the frequency-dependent surface resistance could be calculated for copper- and GNR-based interconnects. Figure 5.7 shows the surface resistance (R_S) of copper and GNR interconnects as a function of frequency.

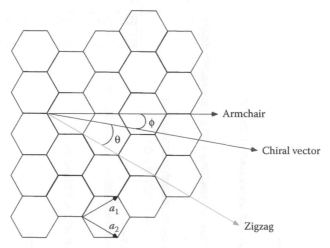

FIGURE 1.3
Schematic of a graphene sheet.

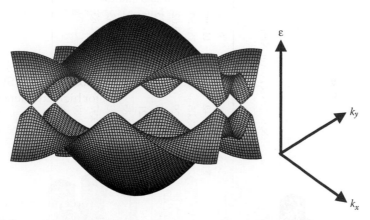

FIGURE 1.6
ε–k relationship of GNR.

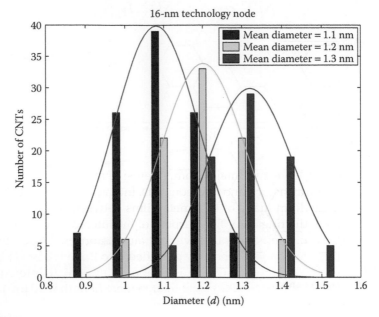

FIGURE 4.2
Distribution of CNTs in a bundle of width 16 nm and thickness 32 nm. The interconnect dimensions are as per 16-nm technology node. The bundle is assumed to be densely packed, that is, with zero spacing.

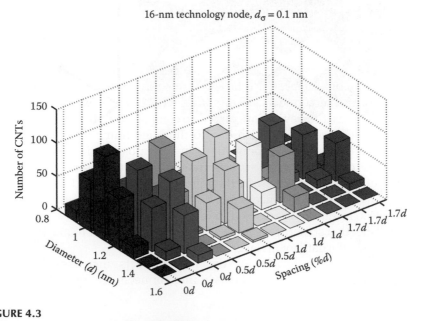

FIGURE 4.3
CNT distribution in a bundle with different CNT diameter and spacing. The spacing is expressed as % of CNT diameter (d) with four different values ($0d$, $0.5d$, $1d$, and $1.7d$). Bundle width, $w = 16$ nm, and thickness, $t = 32$ nm.

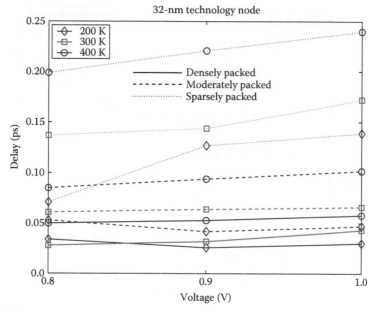

FIGURE 4.19
Propagation delay through CNT bundle at different PTV. The interconnect length is 10 μm and technology node is 32 nm.

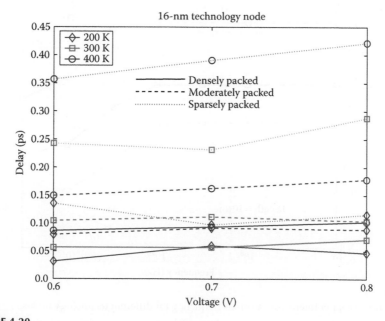

FIGURE 4.20
Propagation delay through CNT bundle at different PTV. The interconnect length is 10 μm and technology node is 16 nm.

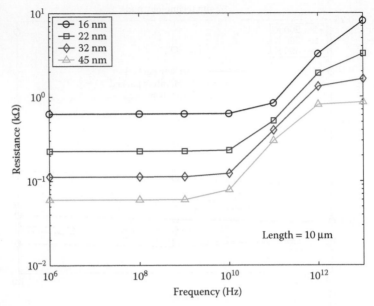

FIGURE 5.1
Resistance of copper interconnect versus frequency for different technology nodes.

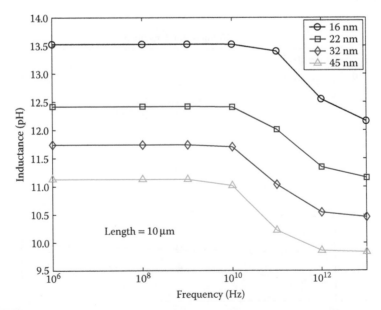

FIGURE 5.2
Inductance of copper interconnect versus frequency for different technology nodes.

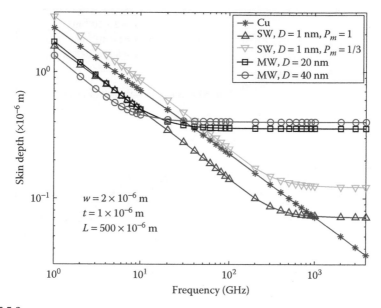

FIGURE 5.3
Skin depth of copper and different CNT interconnect as a function of frequency. PM is the metallic fraction in CNT bundle.

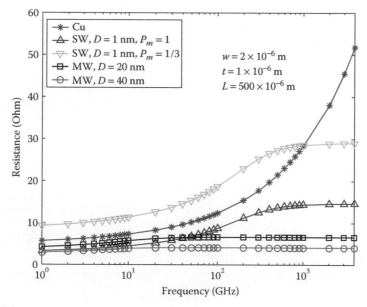

FIGURE 5.4
Resistance of copper and different types of CNT-based interconnect considering skin effect. PM is the metallic fraction in CNT bundle.

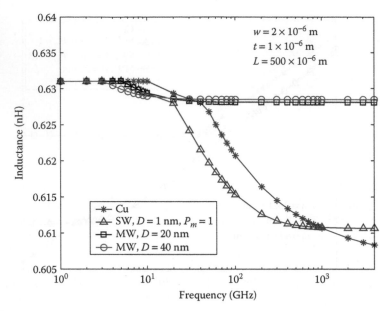

FIGURE 5.5
Inductance of copper and different types of CNT-based interconnect considering skin effect. PM is the metallic fraction in CNT bundle.

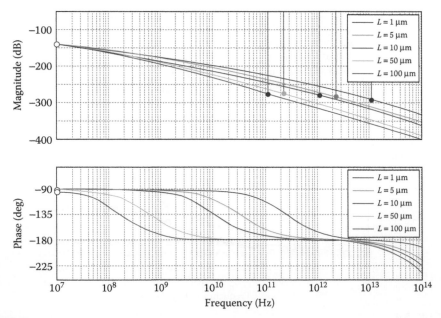

FIGURE 5.20
Bode plots for SWCNT bundle with densely packed CNTs.

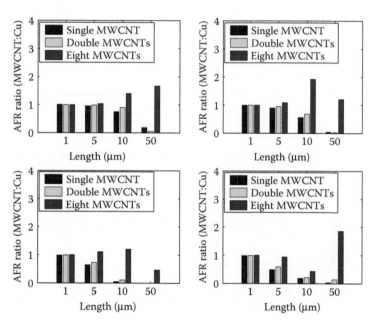

FIGURE 6.7
Ratio of AFR for MWCNT-based interconnect to that of copper-based interconnect (a) 45-nm technology node (b) 32-nm technology node (c) 22-nm technology node (d) 16-nm technology node.

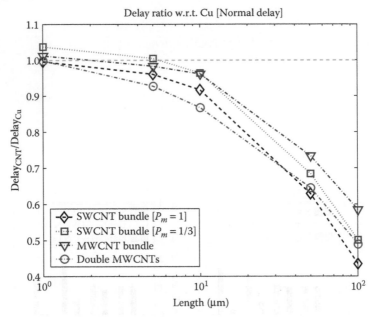

FIGURE 6.8
Normal delay ratio of CNT- and copper-based interconnect as a function of length. Rise and fall delays are averaged.

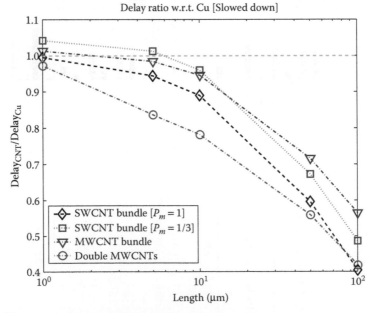

FIGURE 6.9
Delayed crosstalk-induced delay ratio of CNT- and copper-based interconnect as a function of length. Rise and fall delays are averaged.

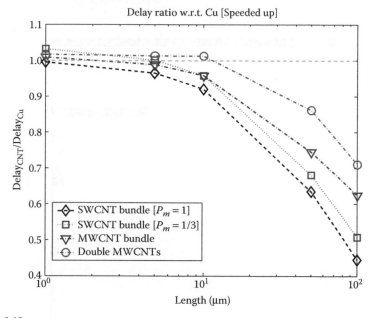

FIGURE 6.10
Speeded crosstalk-induced delay ratio of CNT- and copper-based interconnect as a function of length. Rise and fall delays are averaged.

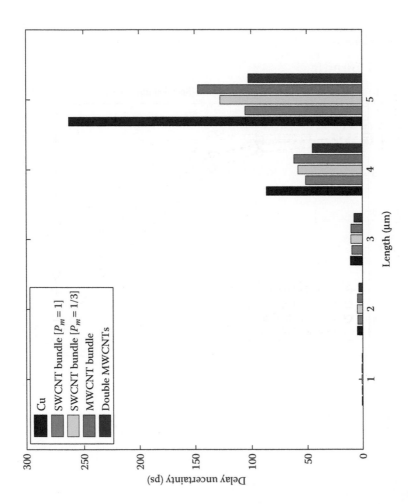

FIGURE 6.11
Delay uncertainty versus interconnect length.

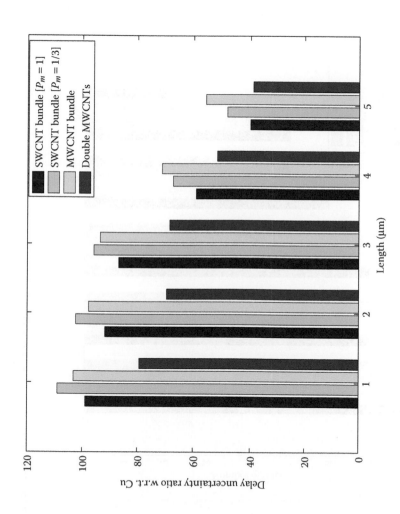

FIGURE 6.12
Delay uncertainty with respect to copper versus interconnect length.

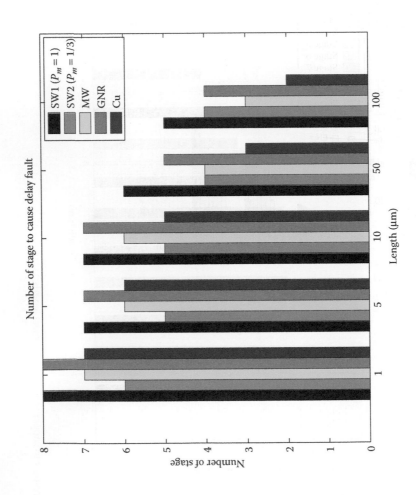

FIGURE 7.25
Number of stage up to which there is no delay-fault versus interconnect length.

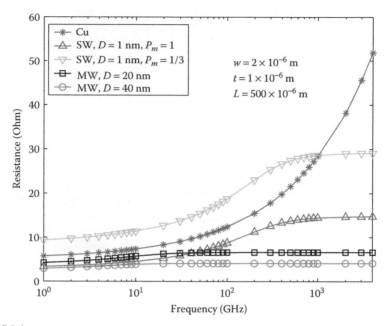

FIGURE 5.4
(**See color insert.**) Resistance of copper and different types of CNT-based interconnect considering skin effect. PM is the metallic fraction in CNT bundle.

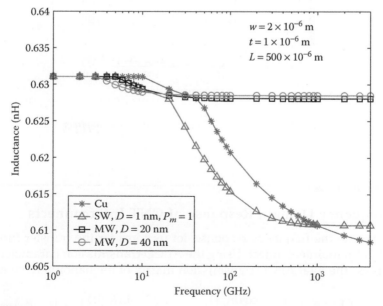

FIGURE 5.5
(**See color insert.**) Inductance of copper and different types of CNT-based interconnect considering skin effect. PM is the metallic fraction in CNT bundle.

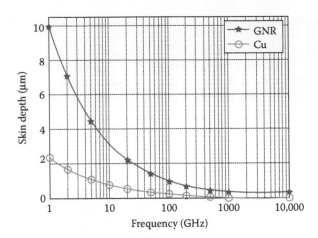

FIGURE 5.6
Variation of skin depth with frequency for copper and GNR.

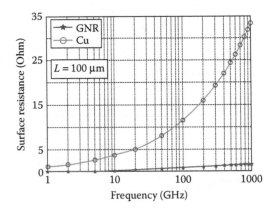

FIGURE 5.7
Resistance of copper- and GNR-based interconnect considering skin effect.

5.3 Frequency Domain Response of CNT Interconnects

To investigate the frequency response, let us consider the transfer function of the CNT nanointerconnect. Using the ABCD transmission parameters, we can express the total ABCD transmission matrix of the nanointerconnect as

$$
T = \begin{pmatrix} 1 & R_t \\ 0 & 1 \end{pmatrix} \times \begin{pmatrix} \cos h(\beta l) & Z_o^{CNT} \sin h(\beta l) \\ \dfrac{1}{Z_o^{CNT}} \sin h(\beta l) & \cos h(\beta l) \end{pmatrix} \times \begin{pmatrix} 1 & R_t \\ 0 & 1 \end{pmatrix} \quad (5.21)
$$

where l = length of CNT interconnect,

$$R_t = \frac{R_C + R_Q}{2} \tag{5.22}$$

$$Z_o^{CNT} = \sqrt{\frac{R + j\omega L}{j\omega C}} \tag{5.23}$$

and

$$\beta = \sqrt{(R + j\omega L) \times (j\omega C)} \tag{5.24}$$

The transfer function of the nanointerconnect system is given by

$$H(j\omega) = \frac{V_{out}(j\omega)}{V_{in}(j\omega)} = \frac{1}{A} \tag{5.25}$$

where A is the first element in the total ABCD matrix. The magnitude of the S_{21} scattering parameter can be calculated by calculating the magnitude of the transfer function $H(j\omega)$ [6].

5.4 RF Simulation Setup

The RF simulation setup is shown in Figure 5.8. Using the methodology discussed in [8] to find out S-parameters [S_{11}, S_{21}], we set up an equivalent circuit model of interconnect based on the extracted resistance, inductance, and capacitance (RLC) parameters explained before.

The CNT transmission line as shown in Figure 5.9 can be simulated in SPICE using the circuit configuration shown in Figure 5.8. The obtained RF transmission characteristics are compared with that obtained from the analytical RF model as explained in previous section. The results are explained in Section 5.5.

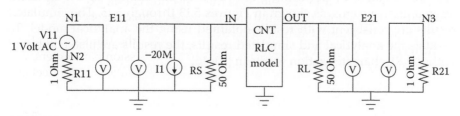

FIGURE 5.8
Circuit configuration to find out scattering parameter of the interconnect system.

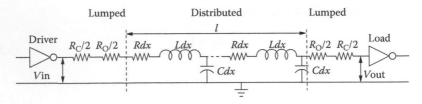

FIGURE 5.9
Transmission line model of CNT-based interconnect.

5.5 RF Simulation Results and Discussions

The RF transmission characteristics of the CNT- and GNR-based interconnects for the future International Technology Roadmap for Semiconductors (ITRS) technology nodes are obtained analytically. The per-unit-length RLC parameters are extracted using the interconnect geometry information obtained from the ITRS [9] interconnect section. MATLAB® can be used for analytical modeling of the RF transmission characteristics.

The RF transmission characteristics obtained through the analytical model of the interconnect systems using SWCNTs and MWCNTs are shown in Figures 5.10 and 5.11. It is observed that as the interconnect length increases, the 3 dB cutoff frequency decreases. For 1-μm-long interconnects, the 3 dB frequency is approximately 100 GHz, whereas for interconnects of length 10 and 100 μm, it becomes approximately 50 and 10 GHz, respectively.

It is also observed that the SWCNT- and MWCNT-based interconnects have almost similar RF responses. The RF performance does not change significantly with the technology scaling from 45 nm down to 16 nm, as observed in Figures 5.10 and 5.11. For a comparative study, the RF transmission characteristics of copper wire are also investigated, as shown in Figure 5.12.

To validate the proposed RF model, the RF transmission characteristics are also investigated, by performing SPICE-based simulation for the different interconnect systems, as shown in Figures 5.13 through 5.15. The simulated results are consistent with results obtained using the analytical model. To validate the analytical and simulated results, the results are also compared with the RF characterization works in [10–16].

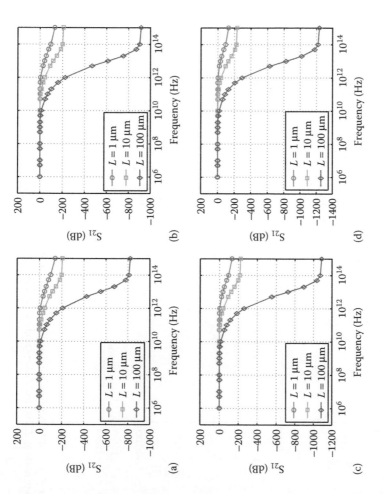

FIGURE 5.10

Transmission coefficient (S_{21}) of SWCNT bundle-based interconnect for different technology nodes obtained from analytical model (a) 45-nm technology node (b) 32-nm technology node (c) 22-nm technology node (d) 16-nm technology node.

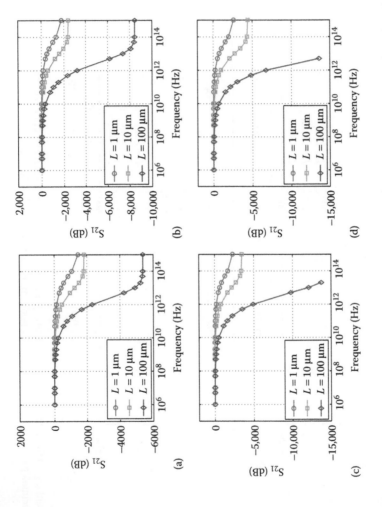

FIGURE 5.11
Transmission coefficient (S_{21}) of MWCNT-based interconnect for different technology nodes, obtained from analytical model (a) 45-nm technology node (b) 32-nm technology node (c) 22-nm technology node (d) 16-nm technology node.

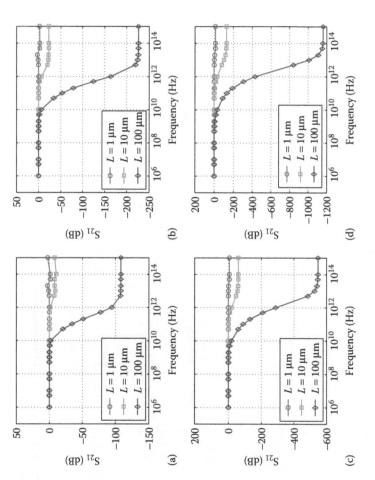

FIGURE 5.12
Transmission coefficient (S_{21}) of copper-based interconnects for different technology nodes, obtained from analytical model (a) 45-nm technology node (b) 32-nm technology node (c) 22-nm technology node (d) 16-nm technology node.

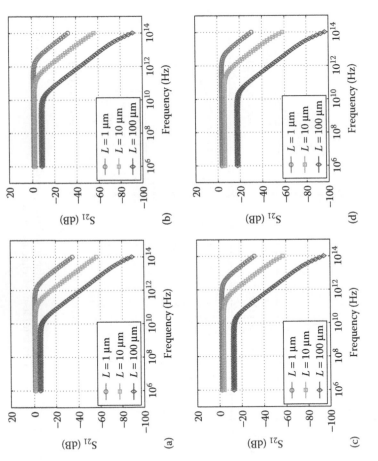

FIGURE 5.13
Transmission coefficient (S_{21}) of SWCNT bundle-based interconnects for different technology nodes, obtained through SPICE simulation (a) 45-nm technology node (b) 32-nm technology node (c) 22-nm technology node (d) 16-nm technology node.

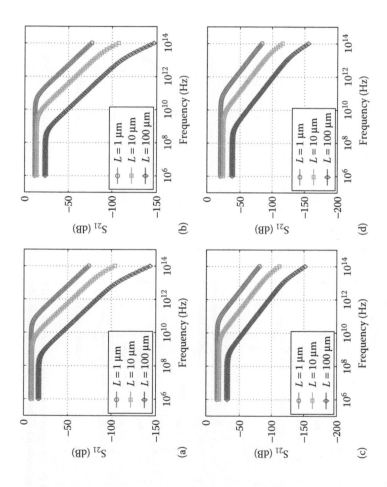

FIGURE 5.14

Transmission coefficient (S_{21}) of MWCNT-based interconnects for different technology nodes, obtained through SPICE simulation (a) 45-nm technology node (b) 32-nm technology node (c) 22-nm technology node (d) 16-nm technology node.

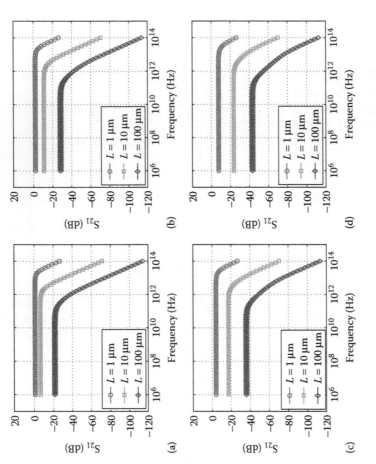

FIGURE 5.15

Transmission coefficient (S_{21}) of copper-based interconnects for different technology nodes, obtained through SPICE simulation (a) 45-nm technology node (b) 32-nm technology node (c) 22-nm technology node (d) 16-nm technology node.

In summary, the following may be concluded: (1) CNT interconnects with shorter lengths (<10 μm) can operate in the THz range; however, for global lengths (>100 μm), the maximum operating frequency is approximately 10 GHz; (2) with technology scaling, RF performance remains almost invariant; (3) SWCNTs and MWCNTs have almost similar RF performance; (4) CNT-based interconnects have an RF performance comparable with that of copper wires.

5.6 Frequency Domain Response of GNR Interconnects

The RF transmission characteristics obtained through analytical model of the GNR interconnect system are shown in Figure 5.16. It is observed that as the interconnect length increases, the 3 dB cutoff frequency decreases. For 1 μm lengths, consistent performance has been observed in and beyond the range of 1000 THz, whereas for interconnect lengths 10 and 100 μm, the cutoff frequency becomes approximately 50 and 10 GHz, respectively. Another interesting observation from the analytical model for 10 μm length gain becomes constant beyond –15 dB.

To validate the proposed RF model, we also investigated RF transmission characteristics, by performing SPICE-based simulations (see Figure 5.17).

The simulated results are consistent with the results obtained using the analytical model. For a comparative study, we also investigated the RF transmission characteristics of copper wire using the model of the Sarkar et al. [17] (Figure 5.18).

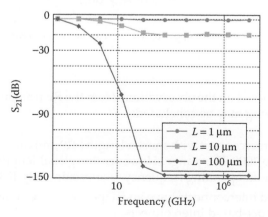

FIGURE 5.16
Transmission coefficient (S_{21}) of GNR-based interconnects for 16-nm technology node, obtained from the analytical model.

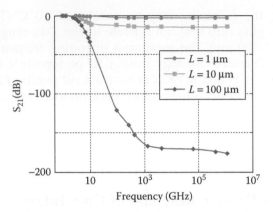

FIGURE 5.17
Transmission coefficient (S_{21}) of GNR-based interconnects for 16-nm technology nodes, obtained through SPICE simulation.

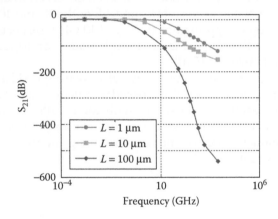

FIGURE 5.18
Transmission coefficient (S_{21}) of copper-based interconnects for 16-nm technology nodes obtained from analytical model.

In summary, based on the observations from the RF transmission model of GNR, the following may be concluded:

1. The GNR interconnect with shorter length (<10 µm) can operate in and beyond the 1000 THz range. However, for global lengths (>100 µm), the maximum operating frequency is approximately 10 GHz.

2. GNR-based interconnects have an RF performance comparable with that of copper-based interconnects.

3. With frequency increase, performance remains almost constant for GNR interconnects for shorter length (<10 µm).

5.7 Stability of CNT and GNR Interconnects

Apart from timing and high-frequency analyses, analysis of the stability of the CNT- and GNR-based interconnect systems is important. The relative stability of a system can be determined using the Nyquist diagrams. The real and imaginary parts of the transfer function are plotted along X and Y axes, respectively, in the Cartesian coordinate system. The frequency is varied to obtain the plot. The point $(-1, 0)$ in the complex plane represents the critical point of stability. Stability is determined by looking at the number of encirclements of the point at $(-1, 0)$. Range of gains over which the system will be stable can be determined by looking at crossing of the real axis.

An alternate method to determine the stability of a system is to use the Bode plots. In Bode plots, the magnitude and phase of the transfer function are plotted as a function of frequency. The gain margin (GM) and phase margin (PM) are calculated from the Bode plots.

5.7.1 Stability Analysis Model of CNT and GNR

A CNT/GNR-based interconnect is modeled as a distributed transmission line connected between a driver and a load, as shown in Figure 5.19.

The transfer function can be written as

$$H(s) = \frac{V_{out}(s)}{V_{in}(s)}$$

$$= \frac{1}{\left[1 + sR_{out}(C_{out} + C_L)\right]\cos h(\theta l) + \begin{bmatrix} R_{out}/Z_0 + sZ_0 C_L \\ + s^2 Z_0 R_{out} C_{out} C_L \end{bmatrix}\sin h(\theta l)} \quad (5.26)$$

where:
$$Z_0 = \sqrt{(R+sL)/(sC)}$$
$$\theta = \sqrt{(R+sL)(sC)}$$
$s = j\omega$ is the complex frequency

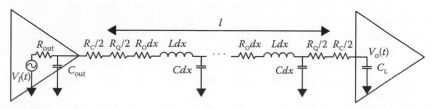

FIGURE 5.19
Schematic of a driver-interconnect-load.

Using the fourth-order Padé's approximation, the transfer function can be written as

$$H(s) = \frac{1}{1 + b_1 s + b_2 s^2 + b_3 s^3 + b_4 s^4} \qquad (5.27)$$

where b_1, b_2, b_3, and b_4 are the parameters that depend on the driver, interconnect, and load.

In [18], Fathi and Forouzandeh showed how the relative stability of the CNT-based interconnect system can be analyzed. They varied the length and diameter of the CNT and found out the relative stability using Nyquist diagrams. They showed that the increase in length and diameter of CNT increases the relative stability of the system.

In [19], Das et al. analyzed the Bode stability of the CNT- and GNR-based interconnect system. Nasiri et al. [20] performed stability analysis in GNR interconnects using the Nyquist stability criterion. They showed that stability of multilayer GNR interconnects can be increased by increasing the length and width of the ribbon.

5.7.2 Results of Stability Analysis in CNT and GNR

In [17,19,21], Das et al. performed the Bode stability analysis. Table 5.1 shows the GM and PM for different interconnect systems. The GM and PM values are calculated from Bode plots, as shown in Figures 5.20 through 5.26.

In the Bode plot, the magnitude (in dB) and phase (in deg) are plotted along the Y axis with respect to frequency along the X axis. Using the MATLAB program, the GM and PM of the interconnect system are calculated. PM is basically the difference between $-180°$ and the phase at 0 dB crosspoint of

TABLE 5.1

Gain Margin and Phase Margin For Different Interconnect Systems

Interconnect Length (μm)→	1		5		10		50		100	
Types of Interconnect ↓	GM	PM	GM	PM	GM	PM	GM	PM	GM	PM
SWCNT bundle densely packed	293	90	283	90	280	90	277	90	277	90
SWCNT bundle sparsely packed	270	90	258	90	253	90	245	90	243	90
Single MWCNT	244	90	232	90	228	90	223	90	222	90
MWCNT bundle of two MWCNTs	244	90	232	90	228	90	223	90	222	90
MWCNT bundle of eight MWCNTs	248	90	238	90	235	90	231	90	231	90
Graphene nanoribbon	250	90	245	90	244	90	243	90	225	90
Copper	285	90	282	90	281	90	279	90	278	90

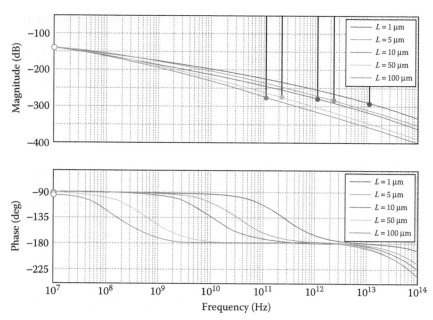

FIGURE 5.20
(See color insert.) Bode plots for SWCNT bundle with densely packed CNTs.

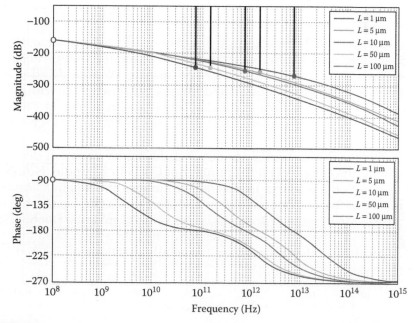

FIGURE 5.21
Bode plots for SWCNT bundle with sparsely packed CNTs.

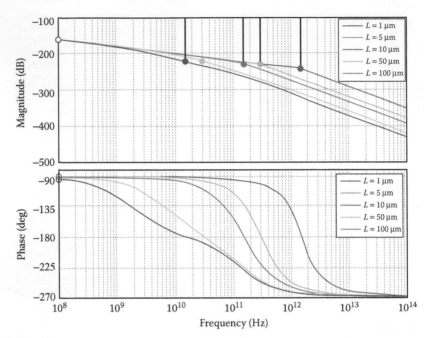

FIGURE 5.22
Bode plot for single MWCNT.

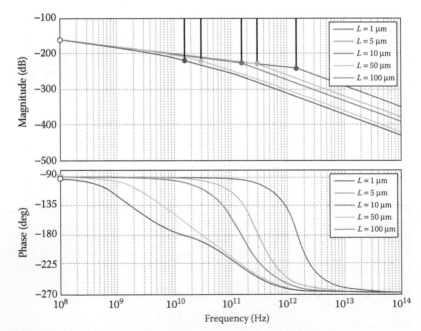

FIGURE 5.23
Bode plot for double MWCNT.

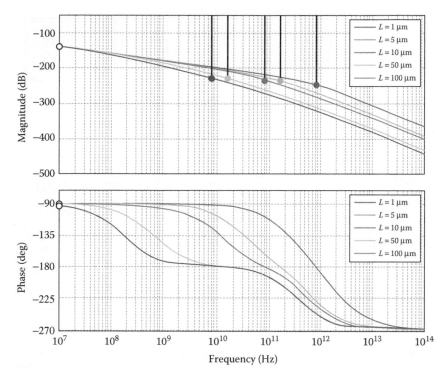

FIGURE 5.24
Bode plot for bundle of eight MWCNT.

the gain curve. GM is just the amount of gain that can be added to move to the 0 dB line at the phase crossover frequency. The system becomes more stable if the GM and PM increase [11].

Let us consider the response of a second-order system with following transfer function:

$$H(s) = \frac{V_{\text{out}}(s)}{V_{\text{in}}(s)} = \frac{1}{LCs^2 + RCs + 1} = \frac{\omega_n^2}{s^2 + 2\zeta\omega_n + \omega_n^2} \qquad (5.28)$$

where:
$\omega_n = 1/\sqrt{LC}$ and $\zeta = R/2\sqrt{C/L}$

To investigate the stability of the system, we must consider three different conditions for stability: (1) whether the system is overdamped, (2) underdamped, or (3) critically damped. Generally damping condition is determined by damping ratio (ζ). If $\zeta > 1$, the system is overdamped; if $\zeta = 1$, the system is critically damped; and if $0 < \zeta < 1$, the system is underdamped.

Consider an underdamped system and an overdamped system with the same undamped natural frequency, but damping ratios ζ_u and ζ_o,

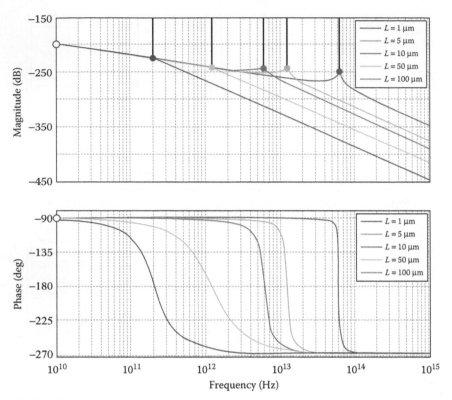

FIGURE 5.25
Bode plot for the GNR interconnect.

respectively. It is shown that the underdamped system is more stable and faster than the overdamped system if and only if [15]

$$\zeta_o < \frac{\zeta_u^2 + 1}{2\zeta_u} \tag{5.29}$$

The stability of the interconnect system is analyzed by increasing its length from 1 to 100 μm. As the length of the interconnect is increased, the switching delay of the RLC network increases. Thus, the system becomes overdamped, is said to be a sluggish system, and reaches the unstable condition. To explain this result further, consider an undamped ($\zeta = 0$) simple oscillator of natural frequency (ω_n). Now let us add damping and increase ζ from 0 to 1.0 [5]. Then, the complex conjugate poles ($-\zeta\omega_n \pm j\omega_d$) will move away from the imaginary axis as ζ increases (because $\zeta\omega_n$ increases), and hence, the level of stability increases [5]. When ζ reaches the value 1.0 (critical damping), we get two identical and real poles at $-\omega_n$. When ζ is increased beyond 1.0, the poles become real and unequal, with one pole having a magnitude smaller than ω_n and the other having a magnitude

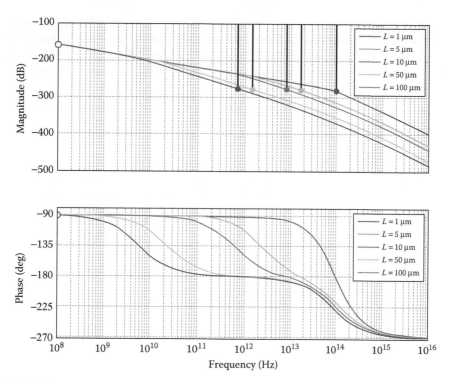

FIGURE 5.26
Bode plot for the copper interconnect.

larger than ω_n. The former (closer to the "origin" of 0) is the dominant pole and will determine both stability and the speed of response of the resulting overdamped system. It follows that as ζ increases beyond 1.0, the two poles will branch out from the location $-\omega_n$, one moving toward the origin (becoming less stable) and the other moving away from the origin [5]. It is now clear that as ζ is increased beyond the point of critical damping, the system becomes less stable. Specifically, for a given value of $\zeta_u < 1.0$, there is a value of $\zeta_o > 1.0$, governed by Equation 5.29, above which the overdamped system is less stable and slower than the underdamped system [5].

5.8 Summary

In this chapter, we showed how high-frequency analysis may be performed on the CNT- and GNR-based interconnects. We showed that CNT- and GNR-based interconnects have the capability of working up to the THz frequency range, leading to extremely fast interconnects. We also presented the stability analysis in CNT and GNR. The relative stability analysis results show that increasing the length and diameter of CNT increases the stability. Similarly, in GNR-based interconnects, increasing the length and width of the GNR increases the stability of the system.

6

Signal Integrity in CNT and GNR Interconnects

6.1 Introduction

In this chapter, we present a study on the crosstalk analysis in the carbon nanotube (CNT) and graphene nanoribbon (GNR) interconnect systems. Crosstalk problems are inevitable in any coupled interconnect system. Due to coupling capacitance between the adjacent interconnects, signal transition on a net in a coupled interconnect system affects the signals on the other nets. The crosstalk-affected net is termed as a *victim net*, and the nets that cause crosstalk on victim nets are termed as *aggressor nets*. Depending on the relative switching of victim and aggressor nets, the crosstalk has several impacts on the circuit performance, as illustrated in Table 6.1.

Considerable research has been made to investigate the applicability of CNT as very large-scale integration (VLSI) interconnect systems. A number of studies on CNT-based interconnects for timing delay, crosstalk noise, and delay analysis have been found [1–5]. Das and Rahaman performed a study to analyze the crosstalk overshoot/undershoot analysis in CNT and GNR interconnects [6–9]. Crosstalk-induced overshoot/undershoots in any form of interconnects impose excess electrical stress on the gate oxide of complementary metal–oxide–semiconductor (CMOS) devices. This excess electrical stress can cause damage to the gate oxide over a period of time and must be kept below a prescribed limit.

As technology is advancing at a rapid rate, gate oxide thickness has shrunk considerably. This causes the vertical electric field across the gate oxide to be extremely high (>7.5 MV/cm). When the device is operated under such a large electric field, the ultrathin gate oxide gets damaged over a period of time. In order to mitigate the increase in electric field across the gate oxide, voltage is also scaled down proportionally to gate oxide thickness. So the gate oxides operated under nominal voltage (V_{DD}) are safe from damage due to high electric field. However, crosstalk-induced overshoot/undershoots force the gate oxides to be operated at a higher voltage than the nominal voltage. This in turn makes the gate oxides more susceptible to damage earlier than

TABLE 6.1

Relative Switching of Aggressor and Victim Nets

Sl. No.	Aggressor Transition	Victim Transition	Results	Impacts
1	Logic 0 → Logic 1	Logic 0 → Logic 1	Decrease in rise time	Timing
2	Logic 0 → Logic 1	Logic 1 → Logic 0	Increase in fall time	Timing
3	Logic 1 → Logic 0	Logic 0 → Logic 1	Increase in rise time	Timing
4	Logic 1 → Logic 0	Logic 1 → Logic 0	Decrease in fall time	Timing
5	Logic 0 → Logic 1	Held at Logic 1	Overshoot	Reliability
6	Logic 0 → Logic 1	Held at Logic 0	Rise glitch	Functionality
7	Logic 1 → Logic 0	Held at Logic 1	Fall glitch	Functionality
8	Logic 1 → Logic 0	Held at Logic 0	Undershoot	Reliability

they would be when gate oxides are operated at nominal voltage. Therefore, it is very important to analyze the effects of crosstalk-induced overshoot/undershoots on the gate oxide reliability of nano-CMOS transistors.

6.2 Crosstalk Analysis in Coupled Interconnect System

In order to analyze the effects of crosstalk on adjacent nets, let us consider the circuit shown in Figure 6.1. The crosstalk-affected net is termed as a victim net and nets that cause crosstalk on a victim net are termed as aggressor nets. Depending on the relative switching of victim and aggressor nets, the crosstalk has several impacts on the circuit performance, as illustrated in Table 6.1.

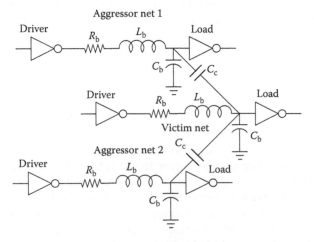

FIGURE 6.1

Schematic of CNT-equivalent circuit to model crosstalk between the adjacent nets. Nets are modeled as L-type RLC network.

As explained in the preceding section, the overshoot/undershoot causes excess electrical stress on the ultrathin gate oxide of the metal–oxide–semiconductor (MOS) devices. For overshoot and undershoot analyses, the victim net is kept fixed at logic levels 1 and 0, respectively. The aggressor nets are switched from logic $0 \rightarrow 1$ and logic $1 \rightarrow 0$.

For modeling CNT-based interconnect systems, both SWCNT and MWCNT are considered. For the SWCNT, a bundle of SWCNTs is used, whereas for MWCNT, only single MWCNTs are used for interconnects. The reason for choosing single MWCNTs for interconnects is due to the fact that the large diameter MWCNTs have been successfully fabricated [10]. For comparison, both L-type and T-type distributed resistance, inductance, and capacitance (RLC) models of interconnects have been considered. The L-type and T-type distributed models are shown in Figures 6.1 and 6.2, respectively.

The impact of the crosstalk-induced overshoot/undershoots on the gate oxide reliability has been estimated in terms of gate oxide failure-in-time (FIT) rate for copper- and CNT-based interconnects. The technology nodes and parameters from the International Technology Roadmap for Semiconductors (ITRS) 2006 road map [11], as illustrated in Table 3.1, have been considered.

6.2.1 Simulation Model

The inverting buffers are designed for different technology nodes for the driver and load in the circuit schematic shown in Figure 6.1. The SPICE models are used from the predictive technology model (PTM) source [12]. The simulations are performed at 45-, 32-, 22-, and 16-nm technology nodes

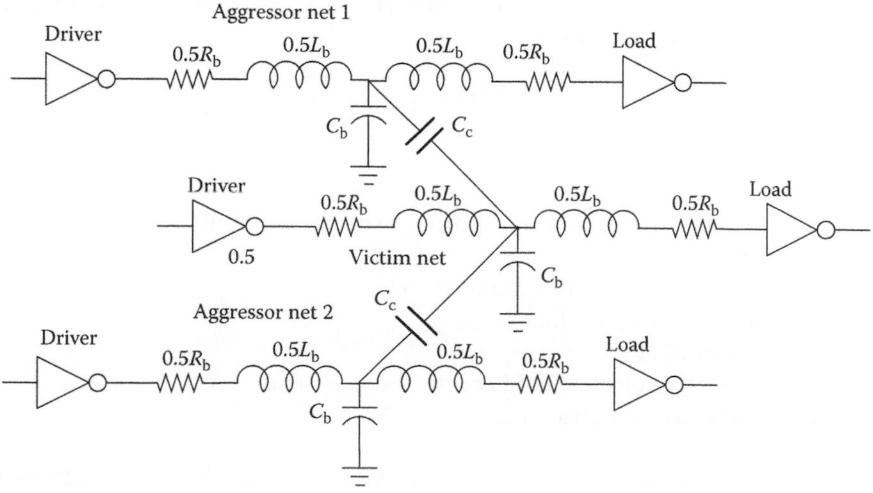

FIGURE 6.2
Schematic of CNT equivalent circuit to model crosstalk between the adjacent nets. Nets are modeled as T-type RLC network.

using Cadence Spectre Circuit Simulator. The effective values of the electric field and average failure rates (AFRs) are calculated by a script written in MATLAB. Simulations are performed for both L-type and T-type distributed networks for interconnects. The results show that there are no significant differences in the results for L-type and T-type network models.

6.3 Gate Oxide Reliability Model

The gate oxides of MOS transistors in a circuit are operated under AC (generally pulse) conditions. Hence, the electric stress on the gate oxide is also an AC electric field. The AC stress can be converted into an equivalent DC stress in order to calculate the effect of the electric field on the gate oxides. The effective electric field across the gate oxide is given by [13]

$$E_{\text{eff}} = \frac{1}{\eta}\left[\frac{1}{T}\int_0^T e^{\eta E_{\text{OX}}(t)}dt\right] \tag{6.1}$$

where η is the field acceleration factor.

In [14], it is shown that the effect of the excess electric field can be estimated by calculating the AFR of the MOS gate oxides. The failure rates of gate oxides are calculated under various electric stress conditions using a reference oxide area and reference temperature. The AFR [13] of an MOS device having gate oxide area A is then expressed as

$$\text{AFR} = \frac{\text{AFR}_{\text{ref}}}{A_{\text{ref}}}Ae^{\eta S_W(E_{\text{eff}}-E_{\text{ref}})} \tag{6.2}$$

where:

A_{ref} is the reference area of the sample used for calculating reference failure rate

AFR_{ref} is the reference failure rate

η is the field acceleration factor

S_W is the Weibull slope

E_{eff} is the effective electric field across the gate oxide

E_{ref} is the reference electric field

Using Equation 6.2, one can calculate the AFR of the MOS devices. The reliability model parameters have been chosen such that the failure rate of a chip operating at nominal voltage is 10 FIT, assuming 1 billion transistors in a chip.

6.4 Results of Gate Oxide Reliability Analysis for CNT-Based Interconnects

The crosstalk-induced overshoot/undershoots are expressed as a percentage of power supply voltage and are plotted as function of interconnect length for different technology nodes. The variation of overshoot/undershoot peak with interconnect length is shown in Figure 6.3.

The width of the overshoot/undershoots is also calculated by calculating the overshoot/undershoot area and approximating the overshoot/undershoot to be triangular in shape. The percentage of overshoot/undershoot width with respect to the pulse width (PW = 2 ns) is shown in Figure 6.4.

The effective electric field across the gate oxide is plotted as a function of interconnect length for different technology nodes as shown in Figure 6.5. The AFR due to the effective electric field is plotted in Figure 6.6.

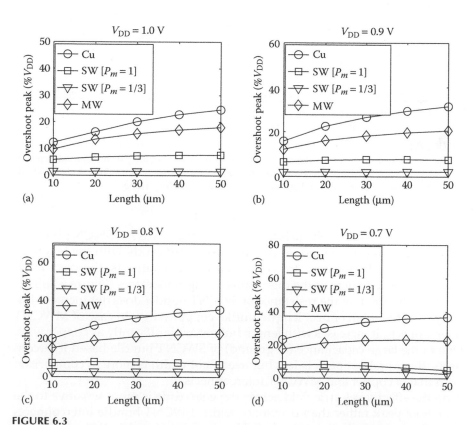

FIGURE 6.3
Crosstalk-induced overshoot peak as a function of interconnect length in different types of interconnects for 45-, 32-, 22-, and 16-nm technology nodes (a) 45-nm technology node (b) 32-nm technology node (c) 22-nm technology node (d) 16-nm technology node.

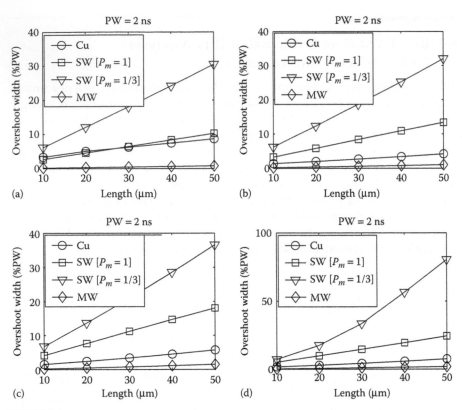

FIGURE 6.4

Crosstalk-induced overshoot width as a function of interconnect length in different types of interconnects for 45-, 32-, 22-, and 16-nm technology nodes (a) 45-nm technology node (b) 32-nm technology node (c) 22-nm technology node (d) 16-nm technology node.

The peak overshoot voltage is below 10% of V_{DD} in case of SWCNT bundle-based interconnect, and it does not increase significantly with the length of the interconnect. However, for copper-based interconnects, the peak overshoot voltage increases as the interconnect length increases (Figure 6.3). This is due to the fact that the resistance of SWCNT bundle does not increase linearly with the interconnect length, unlike copper wires. In case of MWCNT interconnects, it increases with length but not as significantly as copper wire. Due to the large capacitance (to ground) of SWCNT bundle-based interconnects, the overshoot peak is less but overshoot width is more in comparison with that of copper and MWCNT interconnects.

As the effective electric field across the gate oxide is more sensitive to the overshoot peak rather than overshoot width, SWCNT bundle interconnects show minimum effective electric field values compared to those of copper and MWCNT interconnects. Figure 6.5 illustrates the effective electric field for copper-, SWCNT bundle-, and MWCNT-based interconnects.

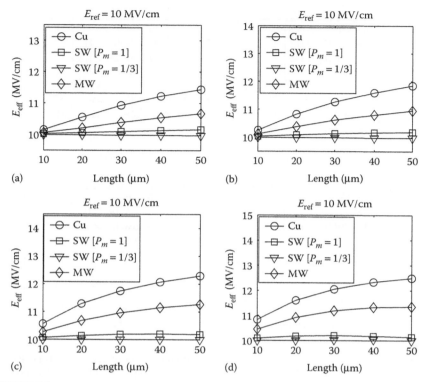

FIGURE 6.5
Effective electric field across the gate oxide versus interconnect length in different types of interconnects (a) 45-nm technology node (b) 32-nm technology node (c) 22-nm technology node (d) 16-nm technology node.

It is observed that the failure rate of the devices increases in orders of magnitude with the increase in length in case of copper-based interconnects (see Figure 6.6), whereas in case of CNT-based interconnect, the increase in AFR is very insignificant. For 45-nm technology node, AFR increases almost 10^5–10^6 orders of magnitude for a fivefold increase in interconnect length in case of copper wires. However, in CNT-based interconnects, the increase is at most three times for SWCNT bundles and 10^2–10^4 times for MWCNT. This indicates that copper-based interconnects are more susceptible to gate oxide damage due to crosstalk overshoot/undershoots. It has also been found that the effect is more pronounced for lower technology nodes in case of copper wires. At the 16-nm technology node, a fivefold increase in length causes an almost 10^6 times increase in AFR in case of copper wires, whereas the increase is just 30% more in case of SWCNT bundle-based interconnects. The sparsely packed SWCNT bundle shows better results as compared to the densely packed bundle. The overshoot peak is 1.5%–6.0% less for the sparsely packed bundle than densely packed, which results in 20%–80% reduction in AFR.

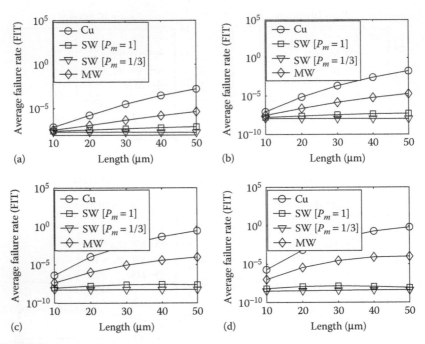

FIGURE 6.6
AFR of buffer circuit versus interconnect length in different types of interconnects (a) 45-nm technology node (b) 32-nm technology node (c) 22-nm technology node (d) 16-nm technology node.

6.5 Analysis for Different Configurations of MWCNT-Based Interconnects

For MWCNT-based interconnects, three different configurations have been used for analysis: (1) single MWCNT of large diameter, (2) double MWCNT of large diameter, and (3) bundle of eight MWCNTs of small diameter.

It is observed that with interconnect length, the overshoot/undershoot parameters (peak, area, and width) increase significantly. This results in increased effective electric field and hence AFR of the MOS devices. It is shown that for smaller lengths (1 μm $\leq l \leq$ 5 μm), the overshoot/undershoot parameters are identical for all four versions of interconnects.

Because AFR is exponentially dependent on the overshoot peak, copper interconnects is more susceptible than MWCNT-based interconnects, and the increase in length of copper wires has a more adverse effect on the gate oxide reliability.

It is observed that the failure rate of the devices increases in orders of magnitude with the increase in length in case of copper-based interconnects, whereas in case of MWCNT-based interconnects, the increase in AFR is insignificant. The failure rate depends on the effective electric field, which in turn critically depends on the peak overshoot voltage.

Typically, the peak overshoot voltage increases with increase in coupling capacitance between the wires, increase in inductance of the wires, increase in resistance of the wires, and decrease in capacitance-to-ground of the wires. In MWCNT-based wires, the coupling capacitance and resistance are less, and capacitance to ground is more as compared to copper-based interconnects. This leads to less peak overshoot voltage in MWCNT-based wires than in copper. However, due to the large kinetic inductance of the MWCNT-based wires, the overshoot peak is greater in MWCNT than that in copper-based wires.

In the case of the 16-nm technology node, AFR increases by almost 5 orders of magnitude—a fivefold increase in interconnect length in case of copper wires. However, for single MWCNT-based interconnects, the increase is 4 orders of magnitude. It is found that copper-based interconnects are more susceptible to gate oxide damage due to crosstalk overshoot/undershoots. Moreover, the effect is further pronounced for lower technology nodes in case of copper wires.

For the 22-nm technology node, a fivefold increase in length causes almost 10^6 times increase in AFR in case of copper wires, whereas the increase is 10^4 times in case of double MWCNT-based interconnects. MWCNT-based interconnects have 10–100 times less AFR across different technology nodes. The failure rate is almost identical for shorter interconnects (1 μm $\leq l \leq$ 5 μm). However, for longer (>10 μm) interconnects, the failure rate is significantly less than that of copper. The bundle-based interconnect is less advantageous than the single or double MWCNT-based interconnects.

Figure 6.7 illustrates the ratio of the AFR of MWCNT-based interconnects to the AFR of copper-based interconnects. With increase in length, MWCNT-based interconnects show better results than copper. However, bundles of MWCNT are not good for gate oxide reliability. Due to less coupling capacitance, single MWCNTs show best results, but this would not meet the predicted aspect ratio of ITRS for different interconnect systems. The double MWCNT-based interconnect shows similar results, and it has less resistance, which will lead to less delay. Therefore, interconnects formed with double MWCNT are best suited for both timing and gate oxide reliability.

6.6 Discussions on Noise and Overshoot/Undershoot Analysis

In the preceding sections, we presented the crosstalk effects in CNT interconnects and their impact on gate oxide reliability. The crosstalk-induced overshoot/undershoots were estimated, and the impact of the overshoot/ undershoots on the gate oxide reliability in terms of FIT rate was calculated.

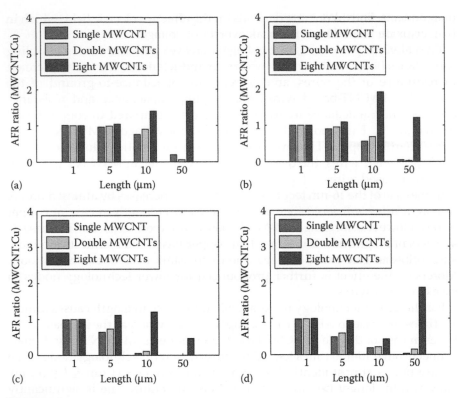

FIGURE 6.7
(See color insert.) Ratio of AFR for MWCNT-based interconnect to that of copper-based interconnect (a) 45-nm technology node (b) 32-nm technology node (c) 22-nm technology node (d) 16-nm technology node.

The comparisons were made with copper-based interconnects. It has been shown that the CNT-based interconnect is more suitable in VLSI circuits as far as gate oxide reliability is concerned. Analysis shows that crosstalk-induced overshoot/undershoots expressed as a percentage of supply voltage remain almost same with the scaling of technology nodes and increase approximately linearly with the length of the copper-based interconnects, whereas in case of CNT-based interconnects, the overshoot/undershoots remain almost invariant with the scaling and interconnect length. It shows that due to the large capacitance-to-ground and small resistance of SWCNT bundle, it has least overshoot/undershoots among three types of interconnects and hence has very small impact on the gate oxide reliability. MWCNT-based interconnects have less impact on the gate oxide reliability compared to copper wire but they are not as good as SWCNT bundle-based interconnects. It has also been shown that sparsely packed SWCNT bundles are better compared to densely packed bundles. Therefore, we can conclude that the impact on gate oxide reliability is not critically influenced with the increase in length of CNT-based interconnects. Both SWCNT- and MWCNT-based interconnects do not impact the gate oxide reliability significantly with technology scaling. As far as gate oxide reliability is concerned, sparsely packed SWCNT bundle-based inter-connects are most suitable in comparison with MWCNT- or copper-based interconnects.

6.7 Analysis of GNR Interconnects

The crosstalk analysis in GNR interconnects has been performed by Das and Rahaman [9] for various lengths (1 μm ≤ l ≤ 1 mm) and the results are com-pared with that of copper and MWCNT interconnects. The results of cross-talk noise analysis are shown in Table 6.2. With the increase in length, the near-end noise peak increases for local and intermediate interconnects, but decreases for global interconnects, whereas the far-end noise peak increases with the interconnect length. For GNR interconnects, the noise peak is smaller as compared to that of copper but greater than MWCNT. This is due to higher inductance and low resistance and capacitance of GNR.

Typically, with increase in interconnect resistance, the near-end noise peak decreases but the far-end noise peak increases, whereas with intercon-nect inductance both increase and with capacitance both decrease. At lower lengths, as the per-unit-length, GNR resistance is higher; the noise peak is less in GNR as compared to longer lengths than that of copper at far-end.

The near-end noise is lesser than the far-end noise due to higher input impedance of load as compared to lower output impedance of the driver. In VLSI circuits, the far-end noise, if greater than the noise threshold, can propagate through the logic gate and can lead to logical errors. In GNR interconnects, the far-end noise is less than that of copper and greater than

TABLE 6.2

Crosstalk Noise in GNR Interconnect for 16-nm Technology Node

V_{DD} = 0.7 V		Near-End			Far-End		
	Length (μm)	Noise Peak (mV)	Noise Area (V.s)	Noise Width (ps)	Noise Peak (mV)	Noise Area (V.s)	Noise Width (ps)
GNR	1	25.5	5.52E-12	432.8	25.8	5.52E-12	427.2
	5	92.6	6.55E-12	141.4	95.7	6.59E-12	137.7
	10	145.6	7.92E-12	108.8	153.3	8.08E-12	105.4
	50	220.4	1.93E-11	175.4	272.1	2.30E-11	169.4
	100	200	3.28E-11	328.4	296.2	4.76E-11	321.3
	500	82.6	1.33E-10	3212	312.5	4.99E-10	3191
	1000	46.4	2.57E-10	11091	314.8	1.71E-09	10846

that of MWCNT-based interconnects for both shorter and longer lengths. However, the noise area is the parameter that determines if the noise can propagate through the logic, which is less in GNR than both Cu and MWCNT interconnects.

Table 6.3 shows the overshoot/undershoot results for GNR interconnect systems. Similar to the noise peak, the overshoot/undershoot peak is also higher in GNR at near-end. However, the far-end overshoot/undershoot in GNR is less as compared to copper. This is again due to the larger per-unit-length inductance and smaller resistance and capacitance of GNR interconnects.

In VLSI circuits, the overshoot/undershoots affect the reliability of the MOS devices. The channel-hot-carrier (CHC) reliability is impacted due to overvoltage at the drain of nMOS transistors rather than at gate [15]. Because the near-end overshoot appears as drain overvoltage, it impacts CHC reliability. On the other hand, the gate oxide reliability is impacted due to the overvoltage at gate of both nMOS and pMOS transistors [14]. As the far-end overshoot/undershoot appears as gate overvoltage, it impacts the gate oxide reliability. GNR shows higher near-end overshoot as compared to that of copper and MWCNT. Therefore, it could be more susceptible to CHC degradation. The far-end peak overshoot is less in GNR than copper but greater than MWCNT. But the far-end overshoot area is less in GNR than both copper and MWCNT. As the gate oxide reliability is more sensitive to the overshoot peak than area, GNR is less susceptible to gate oxide reliability as compared to copper but not as good as MWCNT.

At near-end, the overshoot/undershoot area is almost equal for GNR and copper but less in MWCNT. However, at far-end, the overshoot/undershoot area is less than that of copper and MWCNT. Therefore, as discussed in this section, the CHC reliability is critically impacted in GNR due to its higher overshoot peak at near-end. But gate oxide reliability is less impacted in GNR due to its lower overshoot area at the far-end. The overshoot/undershoot at the

TABLE 6.3

Crosstalk Overshoot in GNR Interconnect for 16-nm Technology Node

$V_{DD} = 0.7$ V		Near-End			Far-End		
	Length (µm)	Overshoot Peak (mV)	Overshoot Area (V.s)	Overshoot Width (ps)	Overshoot Peak (mV)	Overshoot Area (V.s)	Overshoot Width (ps)
GNR	1	26.4	2.62E-13	19.8	26.7	2.65E-13	19.8
	5	86.6	1.21E-12	27.9	89.8	1.25E-12	27.9
	10	127.2	2.32E-12	36.5	135.6	2.48E-12	36.6
	50	177.9	1.13E-11	126.7	241.9	1.50E-11	123.9
	100	160.7	2.30E-11	285.7	272.9	3.77E-11	276.3
	500	75.7	1.24E-10	3276	310	4.90E-10	3160
	1000	45.7	2.54E-10	11102	314.7	1.70E-09	10824

TABLE 6.4

Crosstalk Overshoot/Undershoot Peak, Width, Effective Electric Field across the Gate Oxide, and Average Failure Rate of the CMOS Inverting Buffer for GNR Interconnect; 16-nm Technology Node

Interconnect	Length (µm)	Overshoot Peak (V)	Overshoot Width (ps)	E_{eff} (MV/cm)	AFR (FIT)
GNR	10	0.836	18.31	10.52	1.69E-7
	20	0.885	27.71	10.87	2.56E-6
	30	0.913	38.11	11.26	4.87E-5
	40	0.930	49.52	11.55	4.52E-4
	50	0.942	61.96	11.77	2.48E-3

far-end appears at the gate terminal of the MOS devices of the load cell. This overshoot/undershoot causes the thin gate oxides to be operated at higher voltage due to the crosstalk-induced transient overshoot/undershoot [7]. Therefore, the impact of overshoot/undershoot on gate oxides is estimated by calculating the gate oxide failure rate and comparing the results with those of copper- and MWCNT-based interconnect systems. A detailed crosstalk analysis and its impact on gate oxide reliability are presented in [7]. Using the same gate oxide failure model [7], we have calculated the FIT rate for graphene-based interconnect for various lengths ($10\ \mu m \le l \le 50\ \mu m$). The results are shown in Table 6.4.

The gate oxide reliability analysis shows that the graphene-based interconnect system has less impact on the gate oxides of MOS devices due to crosstalk-induced overshoot/undershoot as compared to that of copper-based interconnects. The FIT rate is 2 orders of magnitude less in graphene-based interconnect compared to copper interconnects. In comparison, MWCNT-based interconnect shows least FIT rate among three different interconnect systems considered in this work. Considering the fact that the GNR fabrication process is much simpler than CNTs, GNR is a better replacement for traditional copper-based interconnect system as far as the gate oxide reliability is concerned.

6.8 Analysis of Delay Uncertainty due to Crosstalk

This section describes how the crosstalk-induced delay is analyzed in CNT and GNR interconnects. Crosstalk introduces a delay uncertainty. The delay uncertainty due to crosstalk in different forms of CNT-based interconnect systems is investigated. As the signal switching speed increases and the density of interconnect grows, the crosstalk phenomena due to the coupling capacitances between interconnects have become a major challenge to the physical designers. The crosstalk has serious effects on the functionality, delay, and reliability of the VLSI circuits. The crosstalk delay problem arises when both aggressor and victim lines switch simultaneously either in same or opposite directions.

The crosstalk effects in SWCNT and MWCNT bus architecture have been analyzed by Rossi et al. [5]. The work in [5] considered triple-wall MWCNT of outer diameter 2 nm for the analysis. Lei and Yin [16] investigated the temperature effects on crosstalk delay and noise in SWCNT and DWCNT interconnects. They used the same interconnect model as in [5].

The work in [2] investigated the crosstalk effects in both SWCNT and DWCNT bundles of dimension corresponding to the ITRS technology nodes by Pu et al. A time and frequency domain model is proposed for crosstalk analysis in SWCNT bundle and MWCNT interconnects in [1,4] by D'Amore et al. In [1], SWCNT bundle-based interconnects of rectangular cross-section are considered but circular cross-sectional interconnects are considered for MWCNT-based interconnects. As the cross-sectional areas are different [1] for two types of CNTs, the results cannot be directly compared.

Sun and Luo [17] proposed a statistical model for analyzing crosstalk noise in SWCNT-based interconnect systems. Two SWCNTs are considered and crosstalk noise induced by the statistical variation of interconnect parameters (CNT diameter, spacing between two tubes, and height from ground plane) due to process variation has been modeled in [17]. Chiariello et al. [3] analyzed frequency-dependent crosstalk noise in CNT bundle-based interconnects considering CNT inductance and resistance. Chen et al. [18] investigated the crosstalk delay for SWCNT-based interconnect systems considering three-line interconnect array.

In [6], the authors presented crosstalk overshoot/undershoot analysis for SWCNT bundle-based interconnects. Though a number of works analyzed crosstalk in CNT-based interconnect systems, none of the previous works have analyzed delay uncertainty due to crosstalk in SWCNT bundle- and MWCNT bundle-based interconnect systems. Previous works showed that delay through SWCNT bundle- and MWCNT-based interconnects are less as compared to traditional copper-based interconnects for intermediate and global wires. However, the crosstalk may speed up or slow down a net depending on the relative switching of aggressor and victim nets, which in turn introduces delay uncertainty. The motivation behind this study is to explore the impact of crosstalk on the delay uncertainty in SWCNT bundle- and MWCNT bundle-based interconnects of dimension corresponding to a future (Year 2019, 16 nm) ITRS technology node.

6.8.1 Crosstalk Delay Analysis in CNT and Copper Interconnects

Crosstalk is a phenomenon that arises due to the coupling capacitance between the parallel interconnects. Figure 6.1 shows the circuit model

TABLE 6.5

Relative Switching of Aggressor and Victim Nets

Aggressor Net 1	Victim Net	Aggressor Net 2	Notation	Effect
Quiet	Rising	Quiet	0 ↑ 0	Normal rise delay
Quiet	Falling	Quiet	0 ↓ 0	Normal fall delay
Rising	Rising	Rising	↑ ↑ ↑	Speedup in rise
Rising	Falling	Rising	↑ ↓ ↓	Slowdown in fall
Falling	Rising	Falling	↓ ↑ ↓	Slowdown in rise
Falling	Falling	Falling	↓ ↓ ↓	Speedup in fall

with one victim and two aggressor nets used for crosstalk analysis. When victim and aggressor nets switch in the same direction, the signal transition becomes faster and the delay through interconnect is decreased (speed up). Alternately, when aggressor nets switch in the opposite direction to victim, the signal transition becomes slower and the delay through the interconnect is increased (slow down). The accuracy of the crosstalk delay analysis depends on interconnect parasitic (RLC) circuit elements and relative switching of the neighboring nets. Crosstalk delay introduces delay uncertainty in the signal transition times, which adversely affects the timing of the VLSI circuits. The relative switching patterns are shown in Table 6.5.

6.8.2 Simulation Results and Discussions

Inverting buffers are designed for the driver and load in the circuit schematic shown in Figure 6.1. The SPICE models are used from the PTM source [12]. The simulations are performed for 16-nm technology node using Cadence Spectre Circuit Simulator and delay values are calculated. It has been assumed that all three nets switch at the same time. For the worst-case delay, two aggressor nets are switched in the opposite direction of victim, whereas for best case, all are switched in the same direction. Aggressor nets are kept quiet for normal delay calculation. Two different SWCNT configurations have been considered: (1) 100% metallic ($P_m = 1$) SWCNT bundle with perfect contact ($R_C = 0$) and (2) 33.3% metallic ($P_m = 1/3$) SWCNT bundle with imperfect contact ($R_C = 100$ kΩ). The MWCNT-based interconnect configurations are considered with perfect contact.

It is observed from the results that the copper-based interconnects are better (delay ratio is greater than unity (Figure 6.8)) for short interconnects. But for longer interconnects (>10 μm), SWCNT and MWCNT bundles show better performance. For instance, sparsely packed SWCNT bundles of length 50 μm show 32.7% reduction in delay as compared to copper, whereas double MWCNT bundles show a 44% reduction. It is observed that as the interconnect length increases, the percentage change in delay due to crosstalk increases for all types of interconnects. The average change in delay due to crosstalk is between 60% and 75% for copper, densely packed SWCNT

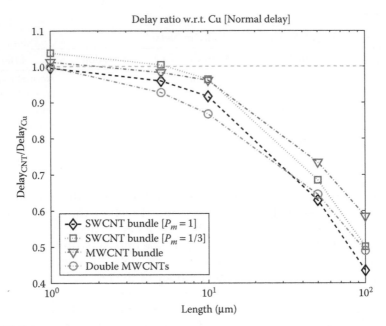

FIGURE 6.8
(See color insert.) Normal delay ratio of CNT- and copper-based interconnect as a function of length. Rise and fall delays are averaged.

bundles, sparsely SWCNT bundles, MWCNT bundles, and double MWCNT bundles. Figures 6.8 through 6.10 show the relative comparison of delay in copper- and CNT-based interconnects.

In Figures 6.8 through 6.10, the delay ratio of 1.0 is the line that decides the relative performance of CNT over copper-based interconnects. When the ratio is greater than unity copper is better and if the ratio is less than unity CNT is better. From Figures 6.8 through 6.10, it is evident that CNT is better for interconnects of length greater than 10 μm. The double MWCNT configuration shows superior performance over copper-based and SWCNT bundle-based interconnects. The crosstalk has also less impact on delay for double MWCNT-based interconnects, which indicates double MWCNT configuration is less susceptible to crosstalk delay than SWCNT/MWCNT bundles.

It is found that as the interconnect length increases, the delay uncertainty of CNT-based interconnects reduces in comparison with that of copper-based interconnects. The 100-μm-long interconnect shows 40%, 48.4%, 55.9%, and 39% delay uncertainty (average value corresponding to rise and fall transition) with respect to copper for densely packed, sparsely packed, and MWCNT bundle-based interconnects, respectively. This result indicates that double MWCNT configuration has least delay uncertainty due to crosstalk among other types of interconnect. Figure 6.11 shows the delay uncertainty for different interconnect systems of length 1 μm $\leq l \leq$ 100 μm.

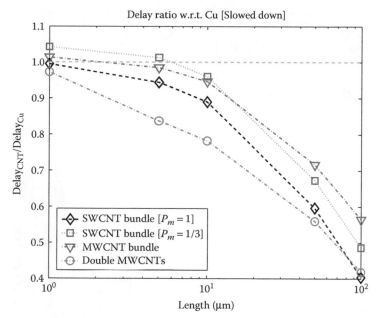

FIGURE 6.9
(See color insert.) Delayed crosstalk-induced delay ratio of CNT- and copper-based interconnect as a function of length. Rise and fall delays are averaged.

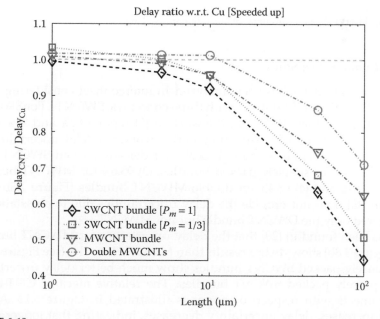

FIGURE 6.10
(See color insert.) Speeded crosstalk-induced delay ratio of CNT- and copper-based interconnect as a function of length. Rise and fall delays are averaged.

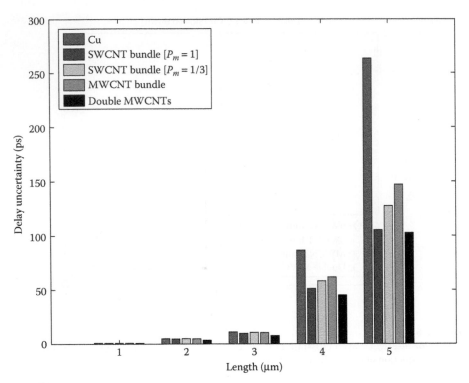

FIGURE 6.11
(See color insert.) Delay uncertainty versus interconnect length.

6.8.3 Comparative Study

In [5], crosstalk delay has been analyzed in interconnect considering three parallel SWCNTs and MWCNT with three concentric SWCNTs. For SWCNTs and MWCNTs of length 10 μm, the work in [5] reports 82% and 94% delay uncertainty without using shielding. In this work, the delay uncertainty for 10-μm-long interconnects show (1) 86.9% for densely packed SWCNT bundles, (2) 95.8% for sparsely packed bundles, (3) 93.6% for MWCNT bundles (Figure 3.8a), and (4) 68.4% for double MWCNT bundles (Figure 3.8b). The work in [19] does not provide the delay uncertainties due to crosstalk as it considered only the DWCNT bundle but not MWCNT.

It has been found in [20] that the delay uncertainty for MWCNT bundles (see Figure 3.8b) shows better results than MWCNT bundles (see Figure 3.8a). The densely packed SWCNT bundles show much better delay uncertainty than sparsely packed SWCNT bundles. The relative merit of CNT-based interconnects with respect to copper is illustrated in Figure 6.12. As the length increases, delay uncertainty decreases, indicating that longer CNT-based interconnects are less susceptible to crosstalk-induced delay. Among four different CNT-based interconnects considered in this work, double

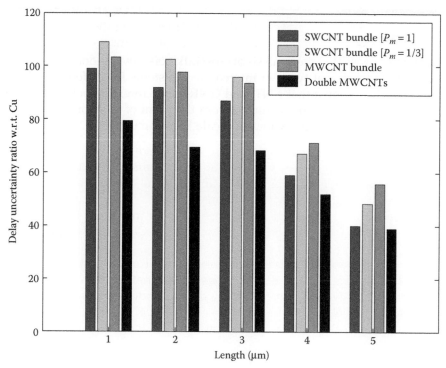

FIGURE 6.12
(See color insert.) Delay uncertainty with respect to copper versus interconnect length.

MWCNT shows the least delay uncertainty. Therefore we may conclude that MWCNT bundles of large diameter are least impacted due to crosstalk delay.

6.8.4 Discussions on Crosstalk Delay Results

In this section, crosstalk delay is analyzed in SWCNT bundle-, MWCNT bundle-, and copper-based interconnect systems for different technology nodes. It is shown that the average crosstalk-induced delay over normal interconnect delay for double MWCNT is better as compared to copper- and SWCNT/MWCNT bundle-based interconnects. The average delay uncertainty with respect to copper interconnects for SWCNT bundle-based interconnects is found to be 75% and 84% for densely and sparsely packed SWCNT bundles, respectively, whereas it is 84% for MWCNT bundle-based interconnects and 62% for double MWCNT-based interconnects. The MWCNT bundle with large diameter MWCNTs is better for crosstalk delay avoidance. The densely packed SWCNT bundle with perfect contact is also suitable for longer interconnects.

6.9 Summary

The chapter presented the analysis of crosstalk noise, overshoot/undershoot, and delay for different types of interconnect systems for different technology nodes. It is shown that CNT- and GNR-based interconnects have great advantage over copper interconnects from the point of view of signal integrity and gate oxide reliability for nanoscale VLSI circuits.

7

Applicability of CNT and GNR as Power Interconnects

7.1 Introduction

In the nanometer regime, the traditional copper-based interconnect system foresees a threat due to its increased resistivity and susceptibility to electromigration. The increased resistivity not only causes increase in delay through signal interconnects, but also poses a serious concern in power interconnects. The higher resistance causes large power supply voltage drop (typically known as IR drop) in the power distribution networks (PDNs), which in turn has adverse effects on signal timing. The fast signal switching and increased inductance of the interconnects (Ldi/dt) put together creates simultaneous switching noise (SSN) on the power nets.

SSN and IR drop are two main problems with the traditional copper-based power nets. They cause fluctuations in the power and ground voltage levels, which result in delay uncertainty and in extreme cases logic failure. In this chapter, we present the results of SSN and IR drop analyses in carbon nanotube (CNT) and graphene nanoribbon (GNR) interconnects and compare their performances with respect to traditional copper-based interconnects.

As the very large scale integration (VLSI) technology advances, the IR drop becomes an important concern in subnanometer designs. The IR drop effect manifests itself in the power and ground distribution networks and has a significant impact on the timing delay of the (VLSI) circuits. The traditional copper-based VLSI interconnects will suffer serious problems beyond 45-nm technology node as predicted by the International Technology Roadmap for Semiconductors (ITRS) [1]. As the width of copper wire decreases, its resistivity increases significantly due to surface roughness and grain boundary scattering [2]. The significant increase in interconnect resistance will lead to increase in IR drop in the power nets, which will lead to performance degradation in VLSI circuits.

CNTs have been proposed as possible replacement for copper-based interconnects in future technologies [3]. Budnik et al. [4] analyzed the performance of SWCNT for on-chip power interconnects. The work in [4] mainly focuses on modeling the resistance of SWCNT-based power interconnects under different bias conditions for different technology nodes. In [5], Naeemi and Meindl modeled CNT-based on-chip power distribution in Gigascale systems. The work in [5] compares SSN and IR drop in copper- and CNT-based interconnects. However, no work has been done on the detailed analysis to estimate the IR drop in realistic conductor dimensions and analyzed its impact on timing. In [6], the authors performed IR drop analysis in CNT-based interconnects and investigated its impact on the timing delay of the VLSI circuits for future ITRS technology nodes. The power and ground interconnects are modeled using both single-walled CNT (SWCNT) and multiwalled CNT (MWCNT).

The equivalent circuit is developed for the IR drop analysis and timing delays are estimated for different types of power interconnect systems. It has been found that the CNT-based PDN has great performance improvement over copper wires as far as the IR drop is concerned for semiglobal and global interconnects.

7.2 IR Drop Analysis in CNT-Based PDN

In a VLSI circuit, the PDN distributes the power supply voltage from the chip power pad to the internal logic cells. The structure of the PDN is often a grid structure. Figure 7.1 illustrates a simplified PDN for a standard-cell-based VLSI design. In this design, standard cells are placed side-by-side in rows. Local power rails between standard cell rows are formed in lower metal layer either Metal 1 or Metal 2, depending on the design cells in the library. The global power rails are always routed in upper metal layer.

Power vias are used to connect the global and local power rails. When there are switching activities on the cells, they will either draw current from the power rails (for charging the output nodes and capacitors) or dump current to the ground rails (for discharging the output nodes). Due to the resistance on the PDN, the current will result in a voltage drop on the power network and/or a voltage increase (also known as ground bounce) on the ground network. IR drop refers to the amount of decrease or increase in the power or ground rail voltage due to the resistance of the power or ground interconnects. IR drop is categorized into two types: static and dynamic IR drop.

Static IR drop is the average voltage drop for the design [7], whereas dynamic IR drop depends on the switching activity of the logic [8], hence is vector dependent. Dynamic IR drop depends on the instantaneous current that is higher while the cell is switching. Dynamic IR drop is caused when large amounts of circuitry switch simultaneously, causing peak current demand [9]. Typically, high IR drop impact on clock networks causes hold-time violations, whereas IR drop on data path signal nets causes setup-time violations. The work in this chapter analyzes the dynamic IR drop in CNT-based PDN.

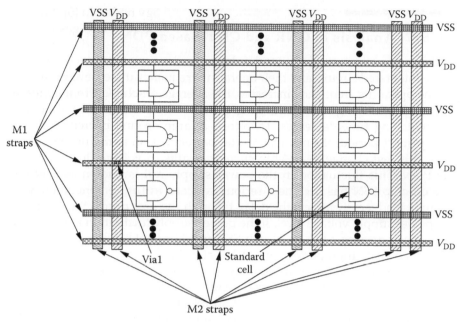

FIGURE 7.1
A PDN for a standard cell-based VLSI design.

In [6], the authors used PDN consisting 10 inverters as standard logic cells placed at a uniform distance. The power and ground rails are modeled by distributed resistance, inductance, and capacitance (RLC) network. For IR drop analysis, they assume that all the 10 cells switch simultaneously. The inverter cells are designed for different technology nodes and the metal–oxide–semiconductor field-effect transistor (MOSFET) models are used from the predictive technology model (PTM) [10]. The cells drive a default load capacitance (Ci) of 0.1 pF, which represents the effective load that the particular cell is driving. The analysis is performed without considering the decoupling capacitances (Cdj). A SPICE netlist of the circuit shown in Figure 7.2 is developed and is simulated in Cadence spectre simulator to find out the dynamic IR drop and the propagation delay of the cells.

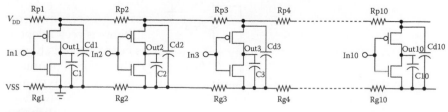

FIGURE 7.2
A resistive power network with 10 inverters connected to the intermediate nodes.

7.3 SSN Analysis in GNR and CNT-Based PDN

With the advancement of VLSI technology, the density of circuit components in the integrated circuits and their operating frequency increase aggressively. As the operating frequency increases, the average on-chip current increases during the charging and discharging of the load capacitor, at the same time the switching time decreases. Therefore, a large on-chip current flows for a very short period of time. Due to the parasitic RLC of the power supply rails, this fast change in on-chip current results in voltage fluctuations in the PDN. This fluctuation in the power supply rail is known as SSN [11]. The on-chip SSN has become a serious concern in future generation VLSI circuits. It adversely affects the delay through the gate due to the change in the power supply voltage level. If the power supply fluctuation is very large, it can also cause logic failure. Hence, SSN must be kept below certain level in the high density high-performance VLSI circuits.

The schematic circuit used for analyzing SSN is shown in Figure 7.3. The logic gate is a complementary metal–oxide–semiconductor (CMOS) inverter that is connected between the power supply rails. The power supply rails are modeled by a lumped RLC network. At the input, a pulse is applied with rise/fall time equal to 1 ps. The CMOS inverter drives an identical inverter of the same size. In Figure 7.3, R_{pwr}, L_{pwr}, and C_{pwr} are the parasitic RLC of the power rail, respectively. Similarly, R_{gnd}, L_{gnd}, and C_{gnd} are the parasitic RLC

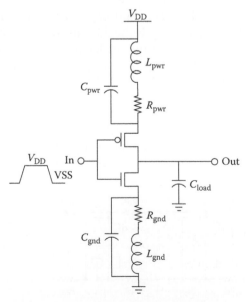

FIGURE 7.3
Schematic circuit of analyzing SSN for a CMOS inverting buffer.

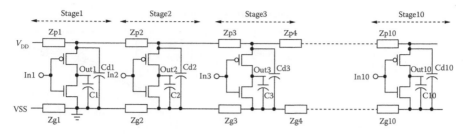

FIGURE 7.4
A RLC power network with 10 inverters connected to the intermediate nodes.

of the ground rail, respectively. The power supply voltage (V_{DD}) is assumed to be 0.7 V [1] and VSS = 0 V.

In [12], the authors used 10 identical CMOS inverters connected to the same power and ground networks to analyze the SSN voltage (Figure 7.4). The length of the power/ground network is varied from 1 to 5 μm. They analyzed SSN without considering any decoupling capacitor between the power and ground rails. The transient IR drop analysis is performed using the same methodology except the RLC interconnect model is replaced by resistive network [11].

The SPICE model is obtained from PTM [10] source. The RLC values of the power/ground rails are calculated for GNR-based power interconnects using the equations presented in Chapter 3. The RLC values for copper wires are obtained from [10]. The circuit is simulated with Cadence spectre simulator. The results are presented in the next section.

7.4 Simulation Results

In this section, we present the results of SSN and IR drop analysis in CNT- and GNR-based PDNs.

7.4.1 IR Drop Analysis Results in CNT Power Nets

The IR drop (ΔV_{DD}) analysis is performed for copper- and CNT-based interconnects of different lengths of the power and ground nets. Four ITRS technology nodes, 45, 32, 22, and 16 nm, are considered for the dynamic IR drop analysis.

By calculating the IR drop at different stages of the inverter, the dynamic IR drop is estimated at different points away from the power and ground pads. The timing delay is then analyzed considering the dynamic IR drop. The peak IR drop values are plotted as a function of stage number of the inverter for different technology nodes as shown in Figures 7.5 through 7.7, for local, semiglobal, and global interconnect lengths for 45-, 32-, 22-, and 16-nm technology nodes, respectively.

Local length ($L = 5$ μm)

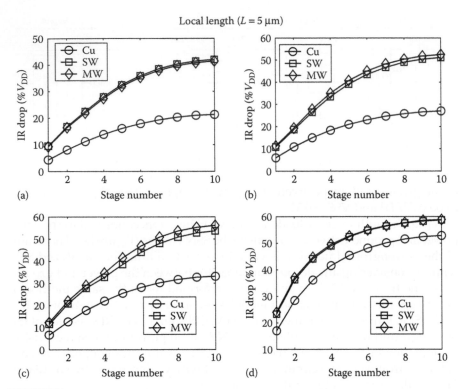

FIGURE 7.5
IR drop (ΔV_{DD}) in local power interconnects for different technology nodes (a) 45-nm technology node (b) 32-nm technology node (c) 22-nm technology node (d) 16-nm technology node.

The separation length indicates how far is a logic circuit from the power and ground pads of the chip. Figures 7.8 through 7.10 show the increase in propagation delay with the increase in dynamic IR drop as progressed away from the power and ground pads of the chip.

For a comparative study, the dynamic IR drop for copper-, SWCNT bundle-, and MWCNT-based power and ground interconnects is plotted as a function of stage number. The cells (inverters) are separated by the uniform distance of three different values: 1, 5, and 10 μm for local lengths; 20, 50, and 100 μm for semiglobal lengths; and 200, 500, and 1000 μm for global lengths. As shown in Figures 7.5 through 7.7, the peak value of the dynamic IR drop value increases as the separation length increases or the stage number increases for three types of interconnect system and for all technology nodes. This is due to the fact that as the separation length or stage number increases, the voltage drop increases successively with the increase in resistance of the wire. For local lengths, copper-based interconnect has least dynamic IR drop. For semiglobal lengths, CNT (both SWCNT and MWCNT)-based interconnect has less dynamic IR drop than copper wires. For global lengths, MWCNT-based interconnect has least dynamic IR drop. CNTs show higher IR drop due to

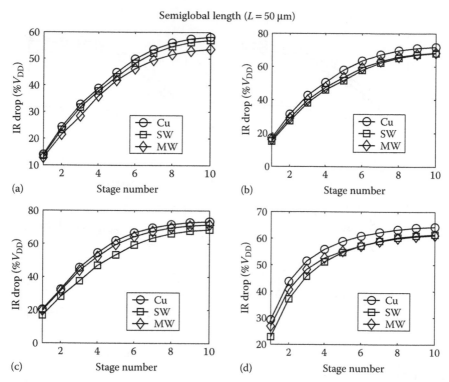

FIGURE 7.6
IR drop (ΔV_{DD}) in semiglobal power interconnects for different technology nodes (a) 45-nm technology node (b) 32-nm technology node (c) 22-nm technology node (d) 16-nm technology node.

the fact that for shorter lengths, CNT has large per-unit-length resistance as compared to copper interconnects.

With the scaling of technology, the IR drop in SWCNT bundle-based interconnect increases for a fixed interconnect length. In 16-nm technology node with 50-μm-long power and ground interconnects, dynamic ΔV_{DD} is 22% less for SWCNT and 8% less for MWCNT as compared to that of copper. For 22-, 32-, and 45-nm technology nodes, they are 18% and 3%, 11% and 6%, and 4% and 9%, respectively. Therefore, with technology scaling SWCNT shows better dynamic ΔV_{DD} for semiglobal lengths. For 500-μm-long power and ground nets, the dynamic ΔV_{DD} is less than copper by 18% and 59%, 30% and 60%, 18% and 52%, and 10% and 49% for 16-, 22-, 32-, and 45-nm technology nodes, respectively. Therefore, MWCNT-based power nets show better dynamic ΔV_{DD} as compared to that of copper and SWCNT for global lengths. The copper-based interconnect shows large increase in dynamic ΔV_{DD} with the increase in separation length as compared to SWCNT- or MWCNT-based interconnects. This can be explained in the following paragraph.

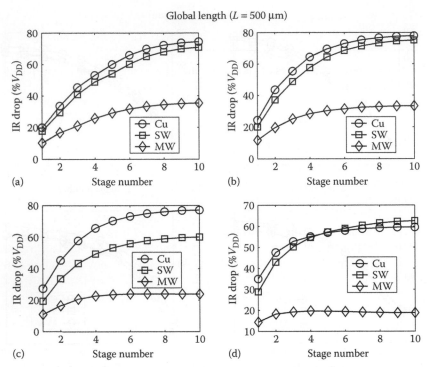

FIGURE 7.7
IR drop (ΔV_{DD}) in global power interconnects for different technology nodes (a) 45-nm technology node (b) 32-nm technology node (c) 22-nm technology node (d) 16-nm technology node..

The per-unit-length resistance of the copper wire is much higher for longer interconnects, which is not the case for both SWCNT- and MWCNT-based interconnects. In CNT-based interconnects, only the ohmic resistance increases with length but the contact resistance and quantum resistance remain constant. Therefore, copper interconnects show greater sensitivity of IR drop to the increase in length. SWCNT bundle has lesser resistance than MWCNT at higher lengths, which makes SWCNT bundle better for semiglobal lengths. However, for global lengths, even though MWCNT has higher resistance than SWCNT bundle, it shows lesser IR drop. This is mainly because of higher capacitance of MWCNT, which acts as decoupling capacitor between the power and ground rails and hence the IR drop is less.

The effects of dynamic IR drop on the timing delay of the circuits are shown in Figures 7.8 through 7.10. It is shown that the delay increases as the separation length increases for the three types of interconnects. However, for global lengths, delay is almost invariant with the separation length for MWCNT. This is due to higher inductance of MWCNT, which makes signal rise/fall times fast. Also due to the large capacitance of MWCNT, the IR drop is less, which makes delay to be insensitive to the separation length. For 50-μm-long power nets, the delay is less in copper for SWCNT and MWCNT by 21% and

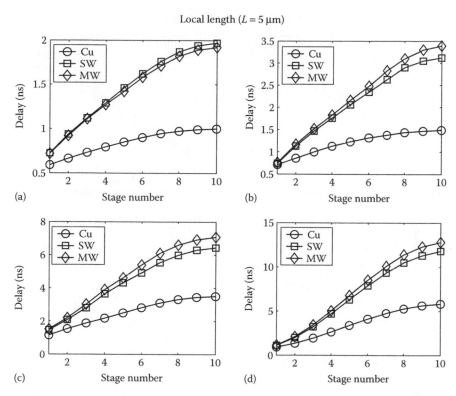

FIGURE 7.8
Timing delay due to dynamic IR drop in local power interconnects for different technology nodes
(a) 45-nm technology node (b) 32-nm technology node (c) 22-nm technology node (d) 16-nm technology node.

11%, 12% and 5%, 10% and 8%, and 2.3% and 6.6% for 16-, 22-, 32-, and 45-nm technology nodes, respectively. For 500-μm-long power nets, the values are 27.5% and 60.3%, −5.8% and 29.5%, 15% and 46.8%, and 10.3% and 25.6%, respectively. As stage number increases, both copper and SWCNT show greater increase in delay; however, it is almost invariant for MWCNT.

7.4.2 Comparison with GNR Power Interconnects

Apart from CNT, graphene-based interconnect is also being explored for the application in VLSI circuits. Graphene has very good electrical and thermal conductivity [13,14]. From the technology point of view, GNR is preferred over CNT due to its better controllability. In this section, we compare our IR drop analysis results with that of GNR power interconnects. Similar to the modeling of CNT, GNR is also modeled by RLC network. The RLC can be estimated using the equations given by Naeemi and Meindl in [15]. To compare the relative merits of CNT with GNR as power interconnects, we analyze IR drop in

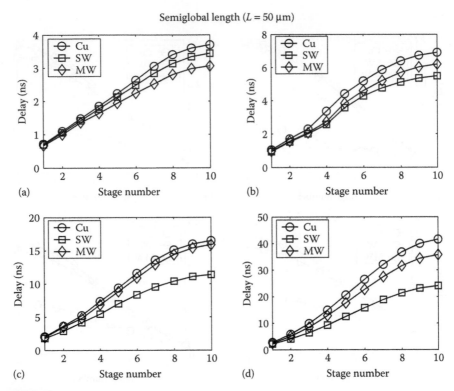

FIGURE 7.9
Timing delay due to dynamic IR drop in semiglobal power interconnects for different technology nodes (a) 45-nm technology node (b) 32-nm technology node (c) 22-nm technology node (d) 16-nm technology node.

GNR-based PDN for 16-nm technology node. In this case, we considered only local interconnects. Figure 7.11 shows the IR drop in GNR power interconnects along with copper- and CNT-based power interconnects. It has been found that GNR shows much better results as compared to both copper- and CNT-based power interconnects due to its very low resistance. The variation in timing delay due to power supply voltage is also calculated, which is shown in Figure 7.12. The delay due to IR drop in GNR power nets is much less as compared to the copper- and CNT-based power nets.

7.4.3 SSN and IR Drop Analyses Results in GNR Power Nets

The peak SSN voltage at different stages in GNR-based PDN for various interconnect length (1 μm ≤ l ≤ 5 μm) is shown in Figure 7.13. Figure 7.14 shows the value of peak SSN in copper-based PDN.

It is found that the GNR-based power interconnects have significantly less SSN than the copper-based power interconnects. For instance, the peak SSN is 412.01 mV in copper at stage 5 for a length of 5 μm, whereas it is only 309.6 mV

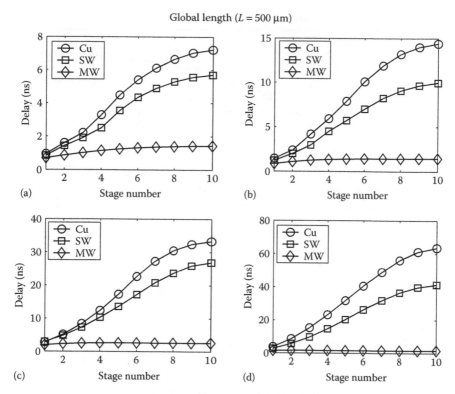

FIGURE 7.10
Timing delay due to dynamic IR drop in global power interconnects for different technology nodes (a) 45-nm technology node (b) 32-nm technology node (c) 22-nm technology node (d) 16-nm technology node.

in GNR-based power interconnect. The percentage reduction of peak SSN in GNR as compared to copper reduces with the increase in interconnect length and with the increase in stage number. Table 7.1 shows the values of peak SSN in GNR and copper for various interconnect length at different stages.

Figures 7.15 and 7.16 show the IR drop in GNR- and copper-based power networks. It is also found that GNR has significantly less IR drop than copper wires. Table 7.2 shows the values of IR drop in GNR- and copper-based power networks. For 5-μm interconnect length at stage 5, the peak IR drop is 507.58 mV in copper and 331.58 mV in GNR. The percentage reduction in peak IR drop reduces with the increase in interconnect length and with the increase in stage number. Due to less resistance of GNR power nets, both the peak SSN and IR drop are significantly less in comparison with that of traditional copper-based PDN.

In order to estimate the impact of the SSN and IR drop on the propagation delay of the CMOS inverter, we calculate the delay at different stages for different interconnect lengths. Figures 7.17 and 7.18 show the delay values for

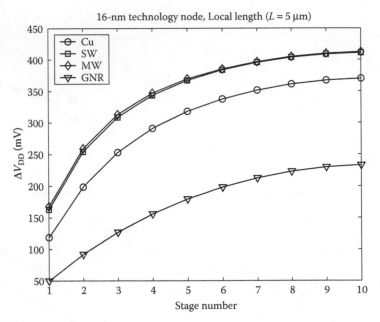

FIGURE 7.11
IR drop (ΔV_{DD}) in different power interconnect system for local lengths in 16-nm technology node.

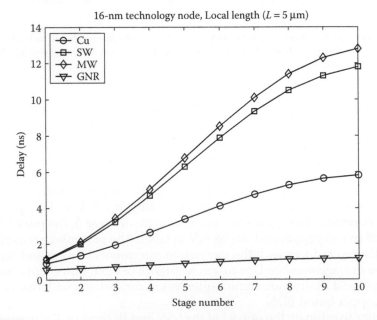

FIGURE 7.12
Delay variations due to IR drop in different power interconnect system for local lengths in 16-nm technology node.

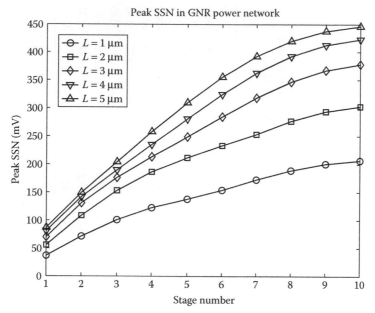

FIGURE 7.13
Peak SSN in GNR power network.

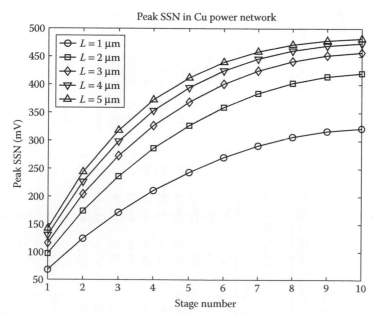

FIGURE 7.14
Peak SSN in copper power network.

TABLE 7.1

Peak SSN in Copper and GNR Power Networks for Different Interconnect Lengths at Different Stages

Length (μm)	Peak SSN in Copper Power Network (mV)					Peak SSN in GNR Power Network (mV)					% Reduction of Peak SSN in GNR w.r.t. Copper				
	1	2	3	4	5	1	2	3	4	5	1	2	3	4	5
Stage 1	67.82	97.15	116.03	130.47	142.17	36.68	55.68	69.89	80.82	86.79	45.91	42.69	39.77	38.05	38.95
Stage 2	124.19	173.95	204.14	226.31	243.66	71.20	108.42	130.76	141.96	149.47	42.67	37.67	35.95	37.27	38.66
Stage 3	171.27	235.82	272.74	298.62	318.05	100.74	153.05	176.17	189.80	204.01	41.18	35.10	35.41	36.44	35.86
Stage 4	210.61	286.14	326.66	353.48	372.61	122.42	186.29	213.16	235.02	257.80	41.87	34.90	34.75	33.51	30.81
Stage 5	243.29	326.93	368.72	394.60	412.01	137.76	211.55	248.67	280.73	309.60	43.38	35.29	32.56	28.86	24.86
Stage 6	269.99	359.39	400.88	424.75	439.81	153.90	233.21	284.43	324.52	355.46	43.00	35.11	29.05	23.60	19.18
Stage 7	291.07	384.40	424.71	446.15	458.79	172.66	253.35	318.26	362.64	392.55	40.68	34.09	25.06	18.72	14.44
Stage 8	306.75	402.58	441.45	460.60	471.17	189.05	276.93	346.81	392.45	419.88	38.37	31.21	21.44	14.80	10.89
Stage 9	317.16	414.44	452.05	469.47	478.55	200.76	293.97	367.30	412.65	437.62	36.70	29.07	18.75	12.10	8.55
Stage 10	322.34	420.29	457.19	473.66	481.98	206.85	302.88	377.96	422.79	446.28	35.83	27.94	17.33	10.74	7.41
					Average % reduction in peak SSN						41	34	29	25	23

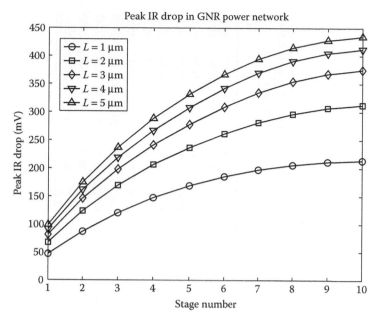

FIGURE 7.15
Peak IR drop in GNR power network.

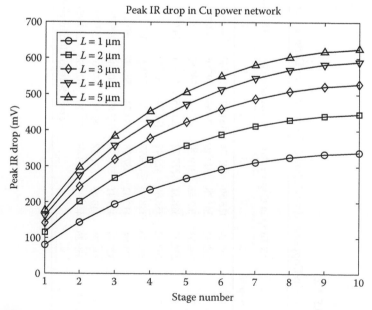

FIGURE 7.16
Peak IR drop in copper power network.

TABLE 7.2

IR Drop in Copper and GNR Power Networks for Different Interconnect Lengths at Different Stages

Length (µm)	IR Drop in Copper Power Network (mV)					IR Drop in GNR Power Network (mV)					% Reduction of Peak IR Drop in GNR w.r.t. Copper				
	1	2	3	4	5	1	2	3	4	5	1	2	3	4	5
Stage 1	80.19	115.59	141.74	161.82	177.42	46.82	67.41	81.20	91.12	98.72	41.61	41.69	42.71	43.69	44.36
Stage 2	143.83	201.78	243.40	274.55	297.98	86.78	123.73	146.10	161.81	174.86	39.66	38.68	39.98	41.06	41.32
Stage 3	194.53	267.42	319.29	357.66	385.87	120.21	169.44	198.16	218.91	236.71	38.20	36.64	37.94	38.79	38.66
Stage 4	234.97	318.31	377.72	421.65	453.61	147.34	206.35	241.18	266.84	288.38	37.29	35.17	36.15	36.72	36.43
Stage 5	267.12	358.06	423.49	472.54	507.58	168.73	236.43	277.09	307.69	331.58	36.83	33.97	34.43	34.89	34.67
Stage 6	292.35	388.95	459.57	513.23	549.99	185.09	261.11	308.75	342.06	366.89	36.69	32.87	32.82	33.35	33.29
Stage 7	311.64	412.42	487.58	544.74	581.97	197.21	281.09	334.43	369.85	394.54	36.72	31.84	31.41	32.11	32.21
Stage 8	325.64	429.41	508.24	567.85	604.46	205.75	296.46	354.31	390.82	414.81	36.82	30.96	30.29	31.18	31.38
Stage 9	334.76	440.48	521.87	582.90	618.69	211.20	306.98	367.88	404.83	428.04	36.91	30.31	29.51	30.55	30.82
Stage 10	339.26	445.95	528.65	590.29	625.56	213.85	312.33	374.77	411.84	434.55	36.97	29.96	29.11	30.23	30.53
Average % reduction in peak IR drop											38	34	34	35	35

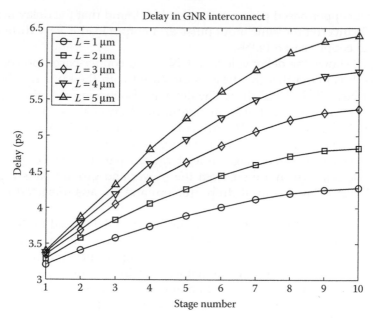

FIGURE 7.17
Delay variation in GNR interconnects.

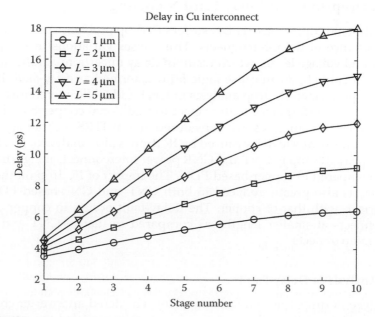

FIGURE 7.18
Delay variation in copper interconnects.

GNR- and copper-based power networks. It is found that the delay through the CMOS inverter for same input pulse and output load is very different in GNR- and copper-based PDNs.

Table 7.3 shows the delay values in GNR- and copper-based power interconnects. It is found that the delay is increased with the stage number due to higher peak SSN and IR drop. The delay is also increased as the length of interconnect is increased. For 5-µm length, the delay at stage 5 is 12.24 ps for copper and 5.24 ps for GNR. The percentage reduction in delay in GNR with respect to copper is increased with the increase in stage number as well as the length of interconnect. For stages 1–10, it varies from 7.07% to 33.4% for 1 µm length. The cells which are far from the power and ground pads typically exhibit larger delay due to IR drop problem and are less impacted in GNR power networks.

Based on the results, we may conclude that the GNR-based PDN is much superior than the traditional copper-based power nets. It has both significantly less SSN and IR drop that are the most important problems in power nets in future generation VLSI circuits. The impact on delay is greatly reduced in GNR in comparison with that of copper. It is also shown that for larger lengths, the impact on delay is much less than that at shorter lengths.

7.5 IR Drop-Induced Delay-Fault Modeling

The traditional copper-based PDN suffers IR drop problem due to the parasitic resistance of the interconnects. This causes variations in the power and ground voltage levels, which result in delay uncertainty. If the delay is significant enough, it can cause logic failure, termed as *delay-fault*. In this work, we performed IR drop analysis in CNT and GNR interconnects and compared the performance with respect to traditional copper-based interconnects. For the analyses, we considered 16-nm ITRS technology node. The logic failure due to delay caused by IR drop is also analyzed. It is found that the peak IR drop in CNT and GNR power interconnects is significantly less as compared to copper-based PDN. The impact of IR drop on the timing delay is also greatly reduced in both CNT- and GNR-based PDNs in comparison with that of copper. The delay-fault occurs in copper power interconnects at smaller lengths as compared to that of CNT and GNR power interconnects.

7.5.1 Analysis of Delay-Fault

To analyze IR drop, Das and Rahaman [16] considered an inverter chain as shown in Figure 7.19. In the circuit diagram, they modeled the power and ground interconnects by series-connected resistive networks to model each

TABLE 7.3

Propagation Delay of CMOS Inverting Buffer due to SSN in Copper and GNR Power Networks for Different Interconnect Lengths at Different Stages

Length (μm)	Delay in Copper Power Network (ps)					Delay in GNR Power Network (ps)					% Reduction of Delay in GNR w.r.t. Copper				
	1	2	3	4	5	1	2	3	4	5	1	2	3	4	5
Stage 1	3.46	3.78	4.08	4.36	4.61	3.21	3.29	3.35	3.38	3.40	7.07	12.90	18.04	22.43	26.33
Stage 2	3.91	4.55	5.20	5.84	6.44	3.41	3.58	3.69	3.79	3.87	12.79	21.49	29.13	35.08	39.91
Stage 3	4.36	5.33	6.34	7.37	8.39	3.58	3.83	4.05	4.19	4.32	17.82	28.12	36.09	43.21	48.52
Stage 4	4.80	6.11	7.48	8.92	10.35	3.74	4.06	4.36	4.61	4.81	22.04	33.56	41.68	48.29	53.57
Stage 5	5.21	6.86	8.59	10.42	12.24	3.89	4.26	4.63	4.95	5.24	25.39	37.87	46.10	52.45	57.18
Stage 6	5.58	7.57	9.62	11.81	13.97	4.01	4.45	4.86	5.25	5.61	28.15	41.15	49.48	55.52	59.86
Stage 7	5.90	8.18	10.52	13.02	15.48	4.12	4.60	5.06	5.50	5.91	30.24	43.76	51.90	57.74	61.83
Stage 8	6.16	8.68	11.25	13.99	16.70	4.20	4.72	5.22	5.70	6.15	31.78	45.62	53.62	59.25	63.18
Stage 9	6.33	9.02	11.75	14.67	17.56	4.25	4.80	5.32	5.83	6.31	32.85	46.84	54.74	60.26	64.05
Stage 10	6.42	9.20	12.01	15.02	18.00	4.28	4.83	5.37	5.89	6.39	33.40	47.48	55.29	60.76	64.49

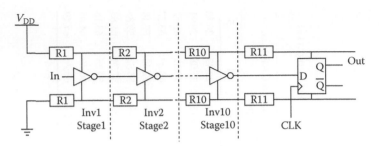

FIGURE 7.19
Schematic circuit for modeling IR drop-induced delay-fault.

segment of power/ground interconnects. All the inverter cells are connected to the same power and ground rails. The connection points are separated by a length l, which is varied from 1 to 100 μm. In [16], the authors took into account the cumulative effects by considering the cascade connection of inverters. By varying the length of the interconnect segment, they analyzed the IR drop at different stages of the inverter chain. To analyze the delay-fault, they connected the output of the inverter chain to the D-input of a D-type flip-flop. If the delay through the inverter chain is significant enough, it can cause a delay-fault, that is, wrong data are stored in the flip-flop. They designed a negative edge-triggered D-type flip-flop using master–slave configuration.

The setup time of the D-type flip-flop is characterized assuming 0.1 ns rise/fall time of the input and clock pulse. A very small capacitance (0.001 fF) is used as the default load. The SPICE model is obtained from PTM for 16-nm technology node. The setup time is found to be 43 ps for the designed D-type flip-flop.

In a D-type flip-flop, if the data input pulse reaches after the setup time of the flip-flop, then the input data cannot be latched. Therefore, the data must reach before the setup time of the flip-flop; otherwise, it can cause a delay-fault. To model the delay-fault, they applied the data input to the flip-flop from the output of the inverter chain.

The input pulse is applied at the input of the first inverter cell. As the data input progresses through the inverter chain, it suffers delay due to the IR drop at different stages of the inverter chain. As long as the delay through the inverter chain is not sufficient to cause the setup time violation, there is no fault in the circuit during the operation. However, if the delay through the inverter chain is large enough, then it may violate the setup time of the flip-flop and can cause a wrong data to be latched at the flip-flop. In [16], the authors varied the length of the power interconnects to investigate the effects of IR drop-induced delay and identify the critical length of power interconnects that can cause a delay-fault.

7.5.2 Simulation Results

The IR drop is analyzed for different interconnect systems in [16]. The variation of IR drop with interconnect length is shown in Figures 7.20 through 7.24 at different stages of the inverter chain. It is observed that as the stage number increases, the IR drop also increases. For 1-µm interconnect length, GNR shows least IR drop as compared to other interconnect systems. It is observed that for the longer lengths, densely packed SWCNT bundle with perfect metallic contact has less IR drop as compared to the other interconnect systems.

The IR drop-induced delay in different interconnect systems is shown in Tables 7.4 through 7.7. The entries in bold indicate setup time violation. It is clear that for longer lengths, setup time violation occurs at smaller stage number. The number of stage (NStage) up to which there is no setup time violation is shown at the bottom of each table. Figure 7.25 shows the plot of NStage versus interconnect length for different interconnect systems.

It is evident from Figure 7.25 that delay-fault occurs in copper power interconnects after second stage for 100-µm length, whereas it occurs after third stage in MWCNT bundle, after fourth stage in sparsely packed SWCNT bundle and GNR, and after fifth stage in densely packed SWCNT bundle. It is also observed that for local lengths (up to 10 µm), GNR and densely packed SWCNT bundle show identical results in delay-fault.

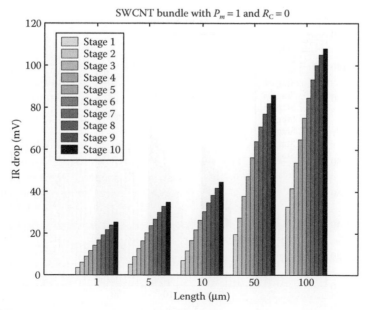

FIGURE 7.20
IR drop in SWCNT bundle (with $P_m = 1$ and $R_C = 0$) interconnect.

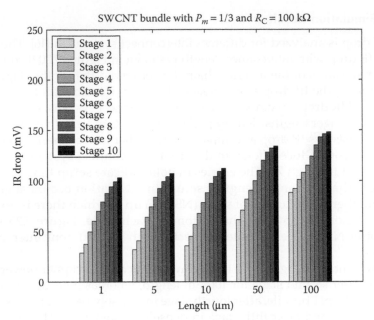

FIGURE 7.21

IR drop in SWCNT bundle (with $P_m = 1/3$ and $R_C = 100$ kΩ) interconnect.

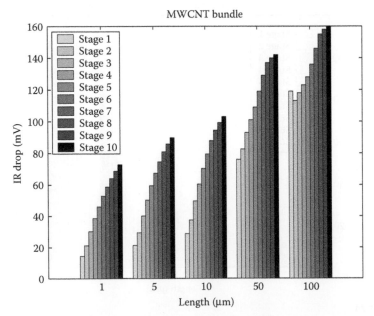

FIGURE 7.22

IR drop in MWCNT bundle interconnect.

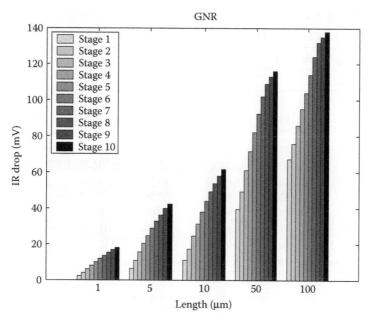

FIGURE 7.23
IR drop in GNR interconnect.

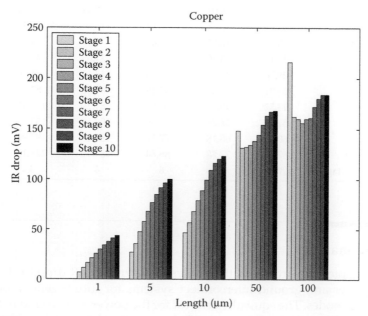

FIGURE 7.24
IR drop in copper interconnect.

TABLE 7.4

IR Drop-Induced Delay in SWCNT Bundle Interconnect

Length (µm)	1	5	10	50	100
Stage			Delay (ps)		
1	5.13	5.19	5.25	5.70	6.19
2	10.80	10.06	11.15	12.62	14.31
3	17.44	17.79	18.21	21.30	24.84
4	24.84	25.46	26.20	31 44	37.45
5	33.14	34.10	35.23	43.29	52.35
6	42.45	43.90	45.54	57.32	70.32
7	52.88	54.98	57.36	74.28	92.73
8	64.45	67.35	70.67	94.43	120.50
9	77.03	80.88	85.28	117.15	152.48
10	88.75	93.56	99.05	139.04	183.60
NStage	6	5	5	4	4

TABLE 7.5

IR Drop-Induced Delay in MWCNT Bundle Interconnect

Length (µm)	1	5	10	50	100
Stage			Delay (ps)		
1	4.91	5.01	5.13	5.93	6.84
2	10.11	10.44	10.81	13.50	16.53
3	15.87	16.62	17.45	23.19	29.54
4	22.11	23.40	24.85	34.65	45.39
5	28.76	30.88	33.16	48.12	64.26
6	35.91	39.04	42.48	64.30	87.15
7	43.52	48.01	52.92	84.21	116.31
8	51.62	57.76	64.50	108.45	153.82
9	60.16	68.18	77.10	136.11	197.92
10	67.85	77.78	88.84	162.94	241.12
NStage	7	6	6	4	3

7.6 Summary

In this chapter, we presented the analysis of dynamic IR drop in the CNT-based power and ground interconnect systems for future generation VLSI technology nodes. The equivalent circuit for the power and ground networks is modeled. The effects of dynamic IR drop on the timing delays have been estimated for different interconnect systems. It is shown that the SWCNT

TABLE 7.6

IR Drop-Induced Delay in GNR Interconnect

Length (µm)	1	5	10	50	100
Stage			Delay (ps)		
1	4.73	4.79	4.86	5.32	5.79
2	9.53	9 73	9.95	11.36	12.96
3	14.57	15.03	15.52	18.67	22.02
4	19.72	20.54	21.43	26.96	32.71
5	24.93	26.25	27.70	36.42	45.21
6	30.21	32.17	34.33	47.32	60.10
7	35.54	38.25	41.30	59.92	78.23
8	40.93	44.56	4S.62	74.26	100.00
9	46.43	51.06	56.28	90.05	124.67
10	50.96	56.62	63.04	105.03	148.52
NStage	8	7	7	5	4

TABLE 7.7

IR Drop-Induced Delay in Copper Interconnect

Length (µm)	1	5	10	50	100
Stage			Delay (ps)		
1	4.79	5.10	5.44	7.51	9.56
2	9.71	10.71	11.78	18.97	26.99
3	15.06	17.24	19.51	34.67	51.71
1	20.60	24.50	28.41	54.09	83.19
5	26.34	32.57	38.65	77.27	121.10
6	32.30	11.60	50.59	105.42	167.18
7	38.45	51.67	64.61	141.61	226.22
8	44.81	62.78	80.81	189.45	306.70
9	51.39	74.82	98.84	246.68	405.61
10	57.01	86.01	116.03	302.95	503.25
NStage	7	6	5	3	2

bundle-based interconnect outperforms the copper- and MWCNT-based wires for semiglobal lengths, whereas MWCNT outperforms copper- and SWCNT-based wires for global lengths. The timing delay increases for both copper- and SWCNT-based systems due to dynamic IR drop, whereas for MWCNT bundle-based systems, the delay increase with the stage number is very insignificant. Hence, we conclude that copper is best suited for local,

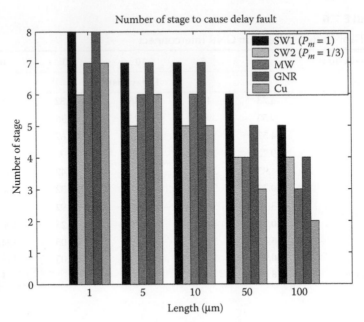

FIGURE 7.25
(See color insert.) Number of stage up to which there is no delay-fault versus interconnect length.

SWCNT is best suited for semiglobal, and MWCNT is best suited for global power nets. The delay-fault is modeled using an inverter chain followed by a D-type. It is observed that the densely packed SWCNT bundle or GNR is better in avoiding delay-fault due to IR drop for local lengths and densely packed SWCNT bundle is better for longer lengths.

8

Future Applications of CNT and GNR

8.1 Introduction

The ever-increasing demand for more circuit components on a single chip requires aggressive scaling of electronic devices and their interconnects. This demand has resulted in many low-dimensional issues. Carbon nanotube (CNT) and more recently graphene have shown exciting electrical properties which can be utilized in making nanoelectronic devices and their interconnects. As the interconnect lateral dimensions are approaching the mean free path of copper (40 nm at room temperature), the grain boundary scattering together with surface scattering leads to rapid increase in the resistivity. Apart from this, the increased susceptibility to electromigration has become a major challenge in nanometric dimensions.

Therefore, the carbon-based nanomaterials CNT and graphene nanoribbon (GNR) may replace the traditional copper interconnects. These nanomaterials show almost a thousand times more current-carrying capacity and significantly higher mean free path than copper.

Several synthesis techniques of CNT and GNR have been explored. The possibility of using them in making circuits also has been explored. The electrical, thermal, and mechanical properties have been studied considering the structural behavior of these materials.

As discussed earlier for interconnect applications, high density aligned CNT bundle is required with precise control over the growth process. Achieving low contact resistance between CNT and other metals is still an open issue.

8.2 Applications of CNT and GNR

Apart from their use in VLSI circuits, CNT and GNR have also been explored for other applications. Recently, CNT arrays on a large area of anodic aluminum oxide (AAO) have been demonstrated as flat panel field emitter with high emitting properties.

The low-voltage field emission display (FED) made with CNT emitters is being explored. The field emission properties of CNT are utilized in making FED.

The cathode ray lightning elements have been fabricated using CNT. It is observed that the luminance of such screens is two times more intense than that of the conventional thermoionic cathode ray tubes. Different colors also have been achieved.

CNT is being used to make scanning tunneling microscope (STM), Atomic force microscopy (AFM), and electrostatic force microscope instruments. Images with very small features are produced with these instruments having CNT-based tips. Biological molecules like DNA can be imaged using nanotube tip STM. AFM nanolithography is now being explored using nanotube tips.

CNTs can be used to make chemical and biological sensors. The electrical resistivity of single-walled CNTs changes in the presence of the gases like NO_2, NH_3, or O_2. The response time of the nanotube-based sensors is 1 order of magnitude faster than that of the conventional sensors.

CNTs are being explored for making nanopipes for precise delivery of liquid and gases. The transport rate of CNT-based nanopipe is also better than that of the conventional microporus materials.

Due to its excellent high strength, CNT is being explored for making high strength, light weight, high-performance composites for making spacecraft and aircraft body parts.

One-dimensional nanowires have been fabricated using CNTs in which foreign materials are used to fill the cavities. A nanothermometer is one application that is made out of metal-filled nanotubes.

CNTs are transparent, flexible, and stretchable. Therefore, they are being explored as conductive layers for touch screen applications. The indium tin oxide (ITO)–based transparent conductors are very expensive and also have very limited flexibility. CNT shows promise as a replacement for ITO.

The low-cost printing over large areas can be achieved using ink or solution of CNTs and graphene. Though the production of CNTs in high volume is yet to be achieved, recent developments by many companies show a great future ahead.

Graphene-based transistors are considered to be potential successors for silicon-based complementary metal–oxide–semiconductor devices. The material is suitable for many high-speed computing applications, potentially enabling terahertz computing, at processor speeds several hundred times faster than silicon.

CNTs and graphene can play an important role in transparent electronics. CNTs, graphene, and their compounds show excellent electrical properties for organic materials. They have a huge potential in electrical and electronic applications such as photovoltaics, sensors, semiconductor devices, displays, conductors, smart textiles, and energy conversion devices (e.g., fuel cells, harvesters, and batteries).

8.3 Summary

Both CNT and GNR have unique properties. Therefore, there are several possibilities in making unique devices using these materials. The advancement of synthesis and growth techniques of these materials will lead to commercial products using CNT and GNR in the near future.

It may be possible to design and manufacture completely new nanoelectronic architecture using these materials. However, a lot needs to be understood and all hidden facts about these materials are to be explored so that these materials can be effectively utilized for making applications that will not be potentially hazardous in future.

References

Chapter 1

1. A. K. Geim and K. S. Novoselov. The rise of graphene. *Nature Materials*, 6(3): 183–191, 2007.
2. H.-P. Boehm, R. Setton, and E. Stumpp. Nomenclature and terminology of graphite intercalation compounds. *Pure and Applied Chemistry*, 66(9):1893–1901, 1994.
3. M. J. O'Connell. *Carbon Nanotube: Properties and Applications*. 1st edition, CRC Press, Boca Raton, FL, 2006.
4. R. Saito, G. Dresselhaus, and M. S. Dresselhaus. *Physical properties of Carbon Nanotubes*. Imperial College Press, London, 1st edition, 1998.
5. P. R. Wallace. The band theory of graphite. *Physical Review*, 71(9):622–634, 1947.
6. Nobelprize.org. The Nobel prize in Physics 2010. http://www.nobelprize.org/ nobel prizes/physics/laureates/2010/, 2010.
7. M. Fujita, K. Wakabayashi, K. Nakada, and K. Kusakabe. Peculiar localized state at zigzag graphite edge. *Journal of the Physical Society of Japan*, 65(7):1920–1923, 1996.
8. A. Javey and J. Kong. *Carbon Nanotube Electronics*, 1st edition. Springer, New York, 2009.
9. R. Saito, G. Dresselhaus, and M. S. Dresselhaus. *Physical Properties of Carbon Nanotubes*, 1st edition. Imperial College Press, London, 1998.

Chapter 2

1. Q. Ngo, D. Petranovic, S. Krishnan, A. M. Cassell, Q. Ye, J. Li, M. Meyyappan, and C. Y. Yang. Electron transport through metal-multiwall carbon nanotube interfaces. *IEEE Transactions on Nanotechnology*, 3(2):311–317, 2004.
2. S. Iijima. Helical microtubules of graphitic carbon. *Nature*, 354:56–58, 1991.
3. L. Delzeit, C. V. Nguyen, B. Chen, R. Stevens, A. Cassell, J. Han, and M. Meyyappan. Multiwalled carbon nanotubes by chemical vapor deposition using multilayered metal catalysts. *Journal of Physical Chemistry B*, 106:5629–5635, 2002.
4. Z. P. Huang, J. W. Xu, Z. F. Ren, J. H. Wang, M. P. Siegal, and P. N. Provencio. Growth of highly oriented carbon nanotubes by plasma-enhanced hot filament chemical vapor deposition. *Applied Physics Letters*, 73(26):3845–3847, 1998.
5. C. Bower, W. Zhu, S. Jin, and O. Zhou. Plasma-induced alignment of carbon nanotubes. *Applied Physics Letters*, 77(6):830–832, 2000.

6. Z. Yao, C. L. Kane, and C. Dekker. High-field electrical transport in single-wall carbon nanotubes. *Physical Review Letters*, 84(13):2941–2944, 2000.

7. L. Delzeit, I. McAninch, B. A. Cruden, D. Hash, B. Chen, J. Han, and M. Meyyappan. Growth of multiwall carbon nanotubes in an inductively coupled plasma reactor. *Journal of Applied Physics*, 91(9):6027–6033, 2002.

8. W. D. Zhang, Y. Wen, J. Li, G. Q. Xu, and L. M. Gan. Synthesis of vertically aligned carbon nanotubes films on silicon wafers by pyrolysis of ethylenediamine. *Thin Solid Films*, 422:120–125, 2002.

9. L. Delzeit, B. Chen, A. Cassell, R. Stevens, C. Nguyen, and M. Meyyappan. Multilayered metal catalysts for controlling the density of single-walled carbon nanotube growth. *Chemical Physics Letters*, 348:368–374, 2001.

10. J. B. Cui, R. Sordan, M. Burghard, and K. Kern. Carbon nanotube memory devices of high charge storage stability. *Applied Physics Letters*, 81(17):3260–3262, 2002.

11. J. Li, Q. Ye, A. Cassell, H. T. Ng, R. Stevens, J. Han, and M. Meyyappan. Bottom-up approach for carbon nanotube interconnects. *Applied Physics Letters*, 82(15):2491–2493, 2003.

12. H. J. Li, W. G. Lu, J. J. Li, X. D. Bai, and C. Z. Gu. Multichannel ballistic transport in multiwall carbon nanotubes. *Physical Review Letters*, 95:086601–086604, 2005.

13. Z. Chen, G. Cao, Z. Lin, I. Koehler, and P. K. Bachmann. A self-assembled synthesis of carbon nanotubes for interconnects. *Nanotechnology*, 17(4):1062–1066, 2006.

14. L. Gomez-Rojas, S. Bhattacharyya, E. Mendoza, D. C. Cox, J. M. Rosolen, and S. R. Silva. RF response of single-walled carbon nanotubes. *Nano Letters*, 7(9):2672–2675, 2007.

15. P. Rice, T. M. Wallis, S. E. Russek, and P. Kabos. Broadband electrical characterization of multiwalled carbon nanotubes and contacts. *Nano Letters*, 7(4):1086–1090, 2007.

16. J. J. Plombon, K. P. O'Brien, F. Gstrein, V. M. Dubin, and Y. Jiao. High-frequency electrical properties of individual and bundled carbon nanotubes. *Applied Physics Letters*, 90:063106, 2007.

17. G. F. Close and H.-S. P. Wong. Assembly and electrical characterization of multiwall carbon nanotube interconnects. *IEEE Transactions on Nanotechnology*, 7(5):596–600, 2008.

18. G. F. Close and H.-S. P. Wong. Measurement of subnanosecond delay through multiwall carbon-nanotube local interconnects in a CMOS integrated circuit. *IEEE Transactions on Electron Devices*, 56(1):43–49, 2009.

19. P. Patel-Predd. Carbon-nanotube wiring gets real. *IEEE Spectrum*, 45(4):14, 2008.

20. W. Wu, S. Krishnan, T. Yamada, X. Sun, P. Wilhite, R. Wu, K. Li, and C. Y. Yang. Contact resistance in carbon nanostructure via interconnects. *Applied Physics Letters*, 94(16):163111-1–163111-3, 2009.

21. A. R. Harutyunyan, G. Chen, T. M. Paronyan, E. M. Pigos, O. A. Kuznetsov, K. Hewaparakrama, S. M. Kim, D. Za-kharov, E. A. Stach, and G. U. Sumanasekera. Preferential growth of single-walled carbon nanotubes with metallic conductivity. *Science*, 326:116–120, 2009.

22. N. Patil, A. Lin, E. R. Myers, K. Ryu, A. Badmaev, C. Zhou, H.-S. P. Wong, and S. Mitra. Wafer-scale growth and transfer of aligned single-walled carbon nanotubes. *IEEE Transactions on Nanotechnology*, 8(4):498–504, 2009.

23. A. D. Franklin and Z. Chen. Length scaling of carbon nanotube transistors. *Nature Nanotechnology*, 5:858–862, 2010.

24. K. Li, R. Wu, P. Wilhite, V. Khera, S. Krishnan, X. Sun, and C. Y. Yang. Extraction of contact resistance in carbon nanofiber via interconnects with varying lengths. *Applied Physics Letters*, 253109-1-253109-3, 2010.

25. X. Chen, D. Akinwande, K.-J. Lee, G. F. Close, S. Yasuda, B. C. Paul, S. Fujita, J. Kong, and H.-S. P. Wong. Fully integrated graphene and carbon nanotube interconnects for giga-hertz high-speed CMOS electronics. *IEEE Transactions on Electron Devices*, 57(11):3137–3143, 2010.

26. Y. Chai, A. Hazeghi, K. Takei, H.-Y. Chen, P. C. H. Chan, A. Javey, and H.-S. P. Wong. Low-resistance electrical contact to carbon nanotubes with graphitic interfacial layer. *IEEE Transactions on Electron Devices*, 59(1):12–19, 2012.

27. J. W. Ward, J. Nichols, T. B. Stachowiak, Q. Ngo, and E. J. Egerton. Reduction of CNT interconnect resistance for the replacement of Cu for future technology nodes. *IEEE Transactions on Nanotechnology*, 11(1):56–62, 2012.

28. P. J. Burke. Luttinger liquid theory as a model of the gigahertz electrical properties of carbon nanotubes. *IEEE Transactions on Nanotechnology*, 1(3):119–144, 2002.

29. P. J. Burke. An RF circuit model for carbon nanotubes. *IEEE Transactions on Nanotechnology*, 2(1):55–58, 2003.

30. S. Salahuddin, M. Lundstrom, and S. Datta. Transport effects on signal propagation in quantum wires. *IEEE Transactions on Electron Devices*, 52(8):1734–1742, 2005.

31. E. Pop, D. Mann, J. Reifenberg, and K. Goodson, H. Dai. Electro-thermal transport in metallic single-wall carbon nanotubes for interconnect applications. *Proceedings of the IEEE International Electron Devices Meeting*, Washington, DC, December 2005, pp. 256–259. Technical Digest of the International Electron Devices Meeting.

32. E. Pop, D. A. Mann, K. E. Goodson, and H. Dai. Electrical and thermal transport in metallic single-wall carbon nanotubes on insulating substrates. *Journal of Applied Physics*, 101(9):093710-1-093710-10, 2007.

33. N. Srivastava and K. Banerjee. Performance analysis of carbon nanotube interconnects for VLSI applications. *Proceedings of the IEEE/ACM International Computer Aided Design Conference*, San Jose, CA, November 2005, pp. 383–390.

34. A. Raychowdhury and K. Roy. Modeling of metallic carbon-nanotube interconnects for circuit simulations and a comparison with Cu interconnects for scaled technologies. *IEEE Transactions on Computer-Aided Design of Integrated Circuits and Systems*, 25(1):58–65, 2006.

35. A. Naeemi and J. D. Meindl. Compact physical models for multiwall carbon-nanotube interconnects. *IEEE Electron Device Letters*, 27(5):338–340, 2006.

36. A. Naeemi and J. D. Meindl. Design and performance modeling for single-walled carbon nanotubes as local, semiglobal, and global interconnects in gigascale integrated systems. *IEEE Transactions on Electron Devices*, 54(1):26–37, 2007.

37. A. Naeemi and J. D. Meindl. Physical modeling of temperature coefficient of resistance for single- and multi-wall carbon nanotube interconnects. *IEEE Electron Device Letters*, 28(2):135–138, 2007.

38. A. Naeemi and J. D. Meindl. Performance modeling for single- and multiwall carbon nanotubes as signal and power interconnects in gigascale systems. *IEEE Transactions on Electron Devices*, 55(10):2574–2582, 2008.

39. Y. Massoud and A. Nieuwoudt. Modeling and design challenges and solutions for carbon nanotube-based interconnect in future high performance integrated circuits. *ACM Journal on Emerging Technologies in Computing Systems*, 2(3):155–196, 2006.

40. A. Nieuwoudt and Y. Massoud. Evaluating the impact of resistance in carbon nanotube bundles for VLSI interconnect using diameter-dependent modeling techniques. *IEEE Transactions on Electron Devices*, 53(10):2460–2466, 2006.

41. A. Nieuwoudt and Y. Massoud. Understanding the impact of inductance in carbon nanotube bundles for VLSI interconnect using scalable modeling techniques. *IEEE Transactions on Nanotechnology*, 5(6):758–765, 2006.

42. A. Nieuwoudt and Y. Massoud. On the impact of process variations for carbon nanotube bundles for VLSI interconnect. *IEEE Transactions on Electron Devices*, 54(3):446–455, 2007.

43. A. Nieuwoudt and Y. Massoud. Performance implications of inductive effects for carbon-nanotube bundle interconnect. *IEEE Electron Device Letters*, 28(4):305–307, 2007.

44. A. Nieuwoudt and Y. Massoud. On the optimal design, performance, and reliability of future carbon nanotube-based interconnect solutions. *IEEE Transactions on Electron Devices*, 55(8):2097–2110, 2008.

45. S. Haruehanroengra and W. Wang. Analyzing conductance of mixed carbon-nanotube bundles for interconnect applications. *IEEE Electron Device Letters*, 28(8):756–759, 2007.

46. D. Rossi, J. M. Cazeaux, C. Metra, F. Lombardi. Modeling crosstalk effects in CNT bus architectures. *IEEE Transactions on Nanotechnology*, 6(2):133–145, 2007.

47. K.-H. Koo, H. Cho, P. Kapur, and K. C. Saraswat. Performance comparisons between carbon nanotubes, optical, and Cu for future high-performance on-chip interconnect applications. *IEEE Transactions on Electron Devices*, 54(12): 3206–3215, 2007.

48. H. Li, W.-Y. Yin, K. Banerjee, and J.-F. Mao. Circuit modeling and performance analysis of multi-walled carbon nanotube interconnects. *IEEE Transactions on Electron Devices*, 55(6):1328–1337, 2008.

49. N. Srivastava, H. Li, F. Kreupl, and K. Banerjee. On the applicability of single-walled carbon nanotubes as VLSI interconnects. *IEEE Transactions on Nanotechnology*, 8(4):542–559, 2009.

50. W. C. Chen, W.-Y. Yin, L. Jia, and Q. H. Liu. Electrothermal characterization of single-walled carbon nanotube (SWCNT) interconnect arrays. *IEEE Transactions on Nanotechnology*, 8(6):718–728, 2009.

51. S.-N. Pu, W.-Y. Yin, J.-F. Mao, and Q. H. Liu. Crosstalk prediction of single- and double-walled carbon-nanotube (SWCNT/DWCNT) bundle interconnects. *IEEE Transactions on Electron Devices*, 56(4):560–568, 2009.

52. H. Li, C. Xu, N. Srivastava, and K. Banerjee. Carbon nanomaterials for next-generation interconnects and passives: Physics, status, and prospects. *IEEE Transactions on Electron Devices*, 56(9):1799–1821, 2009.

53. H. Li and K. Banerjee. High-frequency analysis of carbon nanotube interconnects and implications for on-chip inductor design. *IEEE Transactions on Electron Devices*, 56(10):2202–2214, 2009.

54. D. Fathi and B. Forouzandeh. A novel approach for stability analysis in carbon nanotube interconnects. *IEEE Electron Device Letters*, 30(5):475–477, 2009.

55. S. H. Nasiri, M. K. Moravvej-Farshi, and R. Faez. Stability analysis in graphene nanoribbon interconnects. *IEEE Electron Device Letters*, 31(12):1458–1460, 2010.

56. M. S. Sarto, A. Tamburrano, and M. D'Amore. New electron-waveguide-based modeling for carbon nanotube interconnects. *IEEE Transactions on Nanotechnology*, 8(2):214–225, 2009.

57. M. S. Sarto and A. Tamburrano. Single-conductor transmission-line model of multiwall carbon nanotubes. *IEEE Transactions on Nanotechnology*, 9(1):82–92, 2010.

58. M. S. Sarto and A. Tamburrano. Comparative analysis of TL models for multilayer graphene nanoribbon and multiwall carbon nanotube interconnects. In IEEE International Symposium on Electromagnetic Compatibility (EMC), Fort Lauderdale, FL, July 2010, pp. 212–217.

59. M. S. Sarto, M. D'Amore, and A. Tamburrano. Fast transient analysis of next-generation interconnects based on carbon nanotubes. *IEEE Transactions on Electromagnetic Compatibility*, 52(2):496–503, 2010.

60. F. J. Kurdahi, S. Pasricha, and N. Dutt. Evaluating carbon nanotube global interconnects for chip multiprocessor applications. *IEEE Transactions on Very Large Scale Integration (VLSI) Systems*, 18(9):1376–1380, 2010.

61. A. Naeemi and J. D. Meindl. Compact physics-based circuit models for graphene nanoribbon interconnects. *IEEE Transactions on Electron Devices*, 56(9):1822–1833, 2009.

62. C. Xu, H. Li, and K. Banerjee. Modeling, analysis, and design of graphene nanoribbon interconnects. *IEEE Transactions on Electron Devices*, 56(8):1567–1578, 2009.

63. K.-J. Lee, M. Qazi, J. Kong, and A. P. Chandrakasan. Low-swing signaling on monolithically integrated global graphene interconnects. *IEEE Transactions on Electron Devices*, 57(12):3418–3425, 2010.

64. D. Sarkar, C. Xu, H. Li, and K. Banerjee. High-frequency behavior of graphene-based interconnects part i: Impedance modeling. *IEEE Transactions on Electron Devices*, 58(3):843–852, 2011.

65. D. Sarkar, C. Xu, H. Li, and K. Banerjee. High-frequency behavior of graphene-based interconnects part ii: Impedance analysis and implications for inductor design. *IEEE Transactions on Electron Devices*, 58(3):853–859, 2011.

66. T. Yu, E.-K. Lee, B. Briggs, B. Nagabhirava, and B. Yu. Bilayer graphene/copper hybrid on-chip interconnect: A reliability study. *IEEE Transactions on Nanotechnology*, 10(4):710–714, 2011.

67. T. Yu, C.-W. Liang, C. Kim, E.-S. Song, and B. Yu. Three-dimensional stacked multilayer graphene interconnects. *IEEE Electron Device Letters*, 32(8):1110–1112, 2011.

68. K.-J. Lee, A. P. Chandrakasan, and J. Kong. Breakdown current density of CVD-grown multilayer graphene interconnects. *IEEE Electron Device Letters*, 32(4):557–559, 2011.

69. W.-Y. Yin, J.-P. Cui, W.-S. Zhao, and J. Hu. Signal transmission analysis of multilayer graphene nano-ribbon (MLGNR) interconnects. *IEEE Transactions on Electromagnetic Compatibility*, 54(1):126–132, 2012.

70. S. Rakheja and A. Naeemi. Graphene nanoribbon spin interconnects for nonlocal spin-torque circuits: Comparison of performance and energy per bit with CMOS interconnects. *IEEE Transactions on Electron Devices*, 59(1):51–59, 2012.

71. L. Wilson. International Technology Roadmap for Semiconductors (ITRS) reports, 2006, http://www.itrs.net/reports.html.

72. A. Javey, J. Guo, Q. Wang, M. Lundstrom, and H. Dai. Ballistic carbon nanotube field-effect transistors. *Nature*, 424:654–657, 2003.

73. T. Durkop, S. A. Getty, E. Cobas, and M. S. Fuhrer. Extraordinary mobility in semiconducting carbon nanotubes. *Nano Letters*, 4(1):35–39, 2004.

74. W. Hoenlein, F. Kreupl, G. S. Duesberg, A. P. Graham, M. Liebau, R. V. Seidel, and E. Unger. Carbon nanotube applications in microelectronics. *IEEE Transactions on Components and Packaging Technologies*, 27(4):629–634, 2004.

75. X. Zhou, J.-Y. Park, S. Huang, J. Liu, and P. L. McEuen. Band structure, phonon scattering, and the performance limit of single-walled carbon nanotube transistors. *Physical Review Letters*, 95:146805-1-146805-4, 2005.

76. J. Guo, S. O. Koswatta, N. Neophytou, and M. Lundstrom. Carbon nanotube field-effect transistors. *International Journal of High Speed Electronics and Systems*, 16(4):897–912, 2006.

77. A. Raychowdhury, A. Keshavarzi, J. Kurtin, V. De, and K. Roy. Carbon nanotube field-effect transistors for high-performance digital circuits-DC analysis and modeling toward optimum transistor structure. *IEEE Transactions on Electron Devices*, 53(11):2711–2717, 2006.

78. H.-S. P. Wong. Stanford University CNFET model. http://nano.stanford.edu/, 2014.

79. J. Deng and H.-S. P. Wong. A compact SPICE model for carbon-nanotube field-effect transistors including nonidealities and its application-Part I: Model of the intrinsic channel region. *IEEE Transactions on Electron Devices*, 54(12):3186–3194, 2007.

80. J. Deng and H.-S. P. Wong. A compact SPICE model for carbon-nanotube field-effect transistors including nonidealities and its application-Part II: Full device model and circuit performance benchmarking. *IEEE Transactions on Electron Devices*, 54(12):3195–3205, 2007.

81. S. Sinha, A. Balijepalli, and Y. Cao. Compact model of carbon nanotube transistor and interconnect. *IEEE Transactions on Electron Devices*, 56(10):2232–2242, 2009.

82. S. Lin, Y. Kim, and F. Lombardi. Design of a CNTFET-based SRAM cell by dual-chirality selection. *IEEE Transactions on Nanotechnology*, 9(1):30–37, 2010.

83. J. Zhang, N. Patil, A. Lin, H.-S. P. Wong, and S. Mitra. Carbon nanotube circuits: Living with imperfections and variations. *Proceedings of the Design Automation and Test in Europe Conference and Exhibition*, Dresden, March 8–12, 2010, pp. 1159–1164.

84. G. F. Close, S. Yasuda, B. Paul, S. Fujita, and H.-S. P. Wong. A 1 Ghz integrated circuit with carbon nanotube interconnects and silicon transistors. *Nano Letters*, 8(2):706–709, 2008.

Chapter 3

1. P. J. Burke. Luttinger liquid theory as a model of the gigahertz electrical properties of carbon nanotubes. *IEEE Transactions on Nanotechnology*, 1(3), 119–144, 2002.

2. A. Naeemi and J. D. Meindl. Compact physical models for multiwall carbon-nanotube interconnects. *IEEE Electron Device Letters*, 27(5), 338–340, 2006.

3. A. Naeemi and J. D. Meindl. Physical modeling of temperature coefficient of resistance for single- and multi-wall carbon nanotube interconnects. *IEEE Electron Device Letters*, 28(2), 135–138, 2007.

4. H. Li, W.-Y. Yin, K. Banerjee, and J.-F. Mao. Circuit modeling and performance analysis of multi-walled carbon nanotube interconnects. *IEEE Transactions on Electron Devices*, 55(6), 1328–1337, 2008.
5. H. J. Li, W. G. Lu, J. J. Li, X. D. Bai, and C. Z. Gu, Multichannel ballistic transport in multiwall carbon nanotubes. *Physical Review Letters*, 95, 086601-4, 2005.
6. S. Sato, M. Nihei, A. Mimura, A. Kawabata, D. Kondo, H. Shioya, T. Iwai, M. Mishma, M. Ohfuti, and Y. Awano. Novel approach to fabricating carbon nanotube via interconnects using size-controlled catalyst nanoparticles. *Proceedings of the Interconnect Technology Conference*, Burlingame, CA, 2006, pp. 230–232.
7. A. Nieuwoudt and Y. Massoud. On the optimal design, performance, and reliability of future carbon nanotube-based interconnect solutions. *IEEE Transactions on Electron Devices*, 55(8), 2097–2110, 2008.
8. A. Nieuwoudt and Y. Massoud. Evaluating the impact of resistance in carbon nanotube bundles for VLSI interconnect using diameter-dependent modeling techniques. *IEEE Transactions on Electron Devices*, 53(10), 2460–2466, 2006.
9. A. Nieuwoudt and Y. Massoud. Understanding the impact of inductance in carbon nanotube bundles for VLSI interconnect using scalable modeling techniques, *IEEE Transactions on Nanotechnology*, 5(6), 758–765, 2006.
10. N. Srivastava, H. Li, F. Kreupl, and K. Banerjee. On the applicability of single-walled carbon nanotubes as VLSI interconnects. *IEEE Transactions on Nanotechnology*, 8(4), 542–559, 2009.
11. A. Naeemi and J. D. Meindl. Design and performance modeling for single-walled carbon nanotubes as local, semiglobal, and global interconnects in gigascale integrated systems. 54(1), 26–37, 2007.
12. C. Xu, H. Li, and K. Banerjee. Modeling, analysis, and design of graphene nano-ribbon interconnects, *IEEE Transactions on Electron Devices*, 56(8), 1567–1578, 2009.
13. J.-P. Cui, W.-S. Zhao, W.-Y. Yin, and J. Hu. Signal transmission analysis of multilayer graphene nano-ribbon (MLGNR) interconnects, *IEEE Transactions on Electromagnetic Compatibility*, 54(1), 126–132, 2012.
14. L. Wilson. International Technology Roadmap for Semiconductors (ITRS) reports, 2006, http://www.itrs.net/reports.html.
15. S. H. Nasiri, M. K. Moravvej-Farshi, and R. Faez. Stability analysis in graphene nanoribbon interconnects, *IEEE Electron Device Letters*, 31(12), 1458–1460, 2010.
16. A. Naeemi and J. D. Meindl. Compact physics-based circuit models for graphene nanoribbon interconnects, *IEEE Transactions on Electron Devices*, 56(9), 1822–1833, 2009.
17. A. Naeemi and J. D. Meindl. Conductance modeling for graphene nanoribbon (GNR) interconnects, *IEEE Electron on Device Letters*, 28(5), 428–431, 2007.
18. Yu (Kevin) Cao. Predictive Technology Model, 2008, http://ptm.asu.edu/.
19. K. Fuchs. Conduction electrons in thin metallic films. *Proceedings of the Cambridge Philosophical Soceity*, 34, 100, 1938.
20. E. H. Sondheimer. The mean free path of electrons in metals. *Advances in Physics*, 1(1), 1–42, 1952.
21. A. F. Mayadas and M. Shatzkes. Electrical-resistivity model for polycrystalline films: The case of arbitrary reflection at external surfaces. *Physical Review B*, 1(4), 1382–1389, 1970.

22. E. B. Rosa and F. W. Grover. Formulas and tables for the calculation of mutual and self-inductance, Government Printing Office, BS Sci. Pap. 169, revised 3rd edn., 1916 with corrections 1948.

Chapter 4

1. W. Steinhögl, G. Schindler, G. Steinlesberger, and M. Engelhardt. Size-dependent resistivity of metallic wires in the mesoscopic range. *Physical Review B*, 66, 075414, 2002.
2. P. Kapur, J. P. McVittie, and K. C. Saraswat. Technology and reliability constrained future copper interconnects—Part I: Resistance modelling. *IEEE Transactions on Electron Devices*, 49(4), 590–597, 2002.
3. A. Naeemi, R. Sarvari, and J. D. Meindl. Performance comparison between carbon nanotube and copper interconnects for gigascale integration (GSI). *IEEE Electron Device Letters*, 26(2), 84–86, 2005.
4. K. Banerjee and N. Srivastava. Are carbon nanotubes the future of VLSI Interconnections? *Proceedings of the 43rd ACM/IEEE Design Automation Conference*, San Francisco, CA, 809–814, 2006.
5. L. Wilson. International Technology Roadmap for Semiconductors (ITRS) reports. http://www.itrs.net/reports.html, 2006.
6. N. Srivastava, H. Li, F. Kreupl, and K. Banerjee. On the applicability of single-walled carbon nanotubes as VLSI interconnects. *IEEE Transactions on Nanotechnology*, 8(4), 542–559, 2009.
7. S. Berber, Y.-K. Kwon, and D. Tománek. Unusually high thermal conductivity of carbon nanotubes. *Physical Review Letters*, 84(20), 4613-4616, 2000.
8. P. J. Burke. An RF circuit model for carbon nanotubes. *IEEE Transactions on Nanotechnology*, 2(1), 55–58, 2003.
9. X. Liu, T. Pichler, M. Knupfer, M. S. Golden, J. Fink, H. Kataura, and Y. Achiba. Detailed analysis of the mean diameter and diameter distribution of single-wall carbon nanotubes from their optical response. *Physical Review B*, 66, 2002.
10. C. Lu and J. Liu. Controlling the diameter of carbon nanotubes in chemical vapor deposition method by carbon feeding. *Journal of Physical Chemistry B*, 110, 20254–20257, 2006.
11. A. Naeemi and J. D. Meindl. Design and performance modeling for single-walled carbon nanotubes as local, semiglobal, and global interconnects in gigascale integrated systems. *IEEE Transactions on Electron Devices*, 54(1), 26–37, 2007.
12. N. Srivastava and K. Banerjee. Performance analysis of carbon nanotube interconnects for VLSI applications. *Proceedings of the IEEE/ACM International on Computer Aided Design*, San Jose, CA, 2005, pp. 383–390.
13. A. Nieuwoudt and Y. Massoud. Evaluating the impact of resistance in carbon nanotube bundles for VLSI interconnect using diameter-dependent modeling techniques. *IEEE Transactions on Electron Devices*, 53(10), 2460–2466, 2006.
14. A. Huczko. Synthesis of aligned carbon nanotubes. *Applied Physics A*, 74, 617–638, 2002.

15. C. L. Cheung, A. Kurtz, H. Park, and C. M. Lieber. Diameter-controlled synthesis of carbon nanotubes. *Journal of Physical Chemistry B*, 106, 2429–2433, 2002.

16. I. Hinkov, J. Grand, M. L. de la Chapelle, S. Farhat, C. D. Scott, P. Nikolaev, V. Pichot, P. Launois, J. Y. Mevellec, and S. Lefrant. Effect of temperature on carbon nanotube diameter and bundle arrangement: Microscopic and macroscopic analysis. *Journal of Applied Physics*, 95(4), 2029–2037, 2004.

17. K. C. Narasimhamurthy and R. P. Paily. Impact of bias voltage on magnetic inductance of carbon nanotube interconnects. *Proceedings of the 22nd International Conference on VLSI Design*, New Delhi, January 5–9, 2009, pp. 505–510.

18. A. Naeemi and J. D. Meindl. Impact of electron-phonon scattering on the performance of carbon nanotube interconnects for GSI. *IEEE Electron Device Letters*, 26(7), 476–478, 2005.

19. M. Budnik, A. Raychowdhury, and K. Roy. Power delivery for nanoscale processors with single wall carbon nanotube interconnects. *The 6th IEEE Conference on Nanotechnology*, 2006, pp. 433–436.

20. E. Pop, D. Mann, J. Cao, Q. Wang, K. Goodson, and H. Dai. Negative differential conductance and hot phonons in suspended nanotube molecular wires. *Physical Review Letters*, 95, 155505-1–1155505-4, 2005.

21. E. Pop, D. A. Mann, K. E. Goodson, and H. Dai. Electrical and thermal transport in metallic single-wall carbon nanotubes on insulating substrates. *Journal of Applied Physics*, 101(9), 093710-1-093710-10, 2007.

22. A. Naeemi and J. D. Meindl. Physical modeling of temperature coefficient of resistance for single- and multi-wall carbon nanotube interconnects. *IEEE Electron Device Letters*, 28(2), 135–138, 2007.

23. F. Liang, G. Wang, and W. Ding. Estimation of time delay and repeater insertion in multiwall carbon nanotube interconnects. *IEEE Transactions on Electron Devices*, 58(8), 2712–2720, 2011.

24. H. Li, N. Srivastava, J.-F. Mao, W.-Y. Yin, and K. Banerjee. Carbon nanotube vias: Does ballistic electron–phonon transport imply improved performance and reliability? *IEEE Transactions on Electron Devices*, 58(8), 2689–2701, 2011.

25. L. Jia and W.-Y. Yin. Temperature effects on crosstalk in carbon nanotube interconnects. *Proceedings of the Asia-Pacific Microwave Conference*, Macau, December 16–20, 2008, pp. 1–4.

26. W. C. Chen, W.-Y. Yin, L. Jia, and Q. H. Liu. Electrothermal characterization of single-walled carbon nanotube (SWCNT) interconnect arrays. *IEEE Transactions on Nanotechnology*, 8(6), 718–728, 2009.

27. P. Sun and R. Luo. Analytical modeling for crosstalk noise induced by process variations among CNT-based interconnects. *Proceedings of the IEEE International Symposium on Electromagnetic Compatibility*, Austin, TX, August 17–21, 2009, pp. 103–107.

28. N. Alam, A. K. Kureshi, M. Hasan, and T. Arslan. Performance comparison and variability analysis of CNT bundle and Cu interconnects. *Proceedings of the International Multimedia, Signal Processing and Communication Technologies*, Aligarh, India, March 14–16, 2009, pp. 169–172.

29. A. Nieuwoudt and Y. Massoud. On the impact of process variations for carbon nanotube bundles for VLSI interconnect. *IEEE Transactions on Electron Devices*, 54(3), 446–455, 2007.

30. T. Hiraoka, S. Bandow, H. Shinohara, and S. Iijima. Control on the diameter of single-walled carbon nanotubes by changing the pressure in floating catalyst CVD. *Carbon*, 44, 1853–1859, 2006.

31. A. Naeemi and J. D. Meindl. Performance modeling for single- and multiwall carbon nanotubes as signal and power interconnects in gigascale systems. *IEEE Transactions on Electron Devices*, 55(10), 2574–2582, 2008.

32. D. Das and H. Rahaman. Timing analysis in carbon nanotube interconnects with process, temperature and voltage variations. *Proceedings of the 1st International Symposium on Electronic System Design*, Bhubaneswar, India, December 20–22, 2010, pp. 27–32.

33. A. S. Roy, D. Das, and H. Rahaman. SWCNT based interconnect modeling using verilog-AMS. *Proceedings of the 18th Annual International Conference on High Performance Computing (HiPC) Student Research Symposium*, Bangalore, India, December 18–21, 2011.

34. D. Das and H. Rahaman. Modeling of single-wall carbon nanotube interconnects for different process, temperature, and voltage conditions and investigating timing delay. *Journal of Computational Electronics*, 11(4), 349–363, 2012.

35. H. Li, W.-Y. Yin, K. Banerjee. and J.-F. Mao. Circuit modeling and performance analysis of multi-walled carbon nanotube interconnects. *IEEE Transactions on Electron Devices*, 55(6), 1328–1337, 2008.

36. A. Nieuwoudt and Y. Massoud. On the optimal design, performance, and reliability of future carbon nanotube-based interconnect solutions. *IEEE Transactions on Electron Devices*, 55(8), 2097–2110, 2008.

37. A. Nieuwoudt and Y. Massoud. Understanding the impact of inductance in carbon nanotube bundles for VLSI interconnect using scalable modeling techniques. *IEEE Transactions on Nanotechnology*. 5(6), 758–765, 2006.

38. A. Naeemi and J. D. Meindl. Compact physical models for multiwall carbon-nanotube interconnects, *IEEE Electron Device Letters*, 27(5), 338–340, 2006.

39. S. Pasricha, F. J. Kurdahi, and N. Dutt. Evaluating carbon nanotube global interconnects for chip multiprocessor applications. *IEEE Transactions on Very Large Scale Integration (VLSI) Systems*, 18(9), 1376–1380, 2010.

40. D. Das, S. Das, and H. Rahaman. Design of 4-bit array multiplier using multiwall carbon nanotube interconnects. *Proceedings of the International Symposium on Electronic System Design*, Kolkata, India, December 19–22, 2012.

Chapter 5

1. A. R. Djordjevic, and T. K. Sarkar. Closed-form formulas for frequency-dependent resistance and inductance per unit length of microstrip and strip transmission lines, *IEEE Transactions on Microwave Theory and Techniques*, 42(2), 241–248, 1994.

2. M. Kamon, M. J. Tsuk, and J. White. FastHenry: A multipole-accelerated 3-D inductance extraction program. *IEEE Transactions on Microwave Theory and Techniques*, 42(9), 1750–1758, 1994.

3. X. Qi, G. Wang, Z. Yu, R. W. Dutton, T. Young, and N. Chang. On-chip inductance modeling and RLC extraction of VLSI interconnects for circuit simulation. *Proceedings of the IEEE Custom Integrated Circuits Conference*, Orlando, FL, May 21–24, 2000, pp. 487–490.

4. N. W. Ashcroft and N. D. Mermin. *Solid State Physics*. Saunders College, Philadelphia, PA, 1976.

5. H. Li and K. Banerjee. High-frequency analysis of carbon nanotube interconnects and implications for on-chip inductor design. *IEEE Transactions on Electron Devices*, 56(10), 2202–2214, 2009.

6. D. M. Pozar. *Microwave Engineering*. Wiley, New York, 1998.

7. D. Sarkar, C. Xu, H. Li, and K. Banerjee. High-frequency behaviour of graphene based interconnects—Part I: Impedance modeling. *IEEE Transactions on Electron Devices*, 58(3), 843–852, 2011.

8. R. Goyal. S-parameter output from the SPICE program, *IEEE Circuits and Devices Magazine*, 4(2), 28–29, 1988.

9. L. Wilson. International Technology Roadmap for Semiconductors (ITRS) reports. http://www.itrs.net/reports.html, 2006.

10. L. Gomez-Rojas, S. Bhattacharyya, E. Mendoza, D. C. Cox, J. M. Rosolen, and S. R. Silva, RF response of single-walled carbon nanotubes. *Nano Letters*, 7(9), 2672–2675, 2007.

11. P. Rice, T. M. Wallis, S. E. Russek, and P. Kabos. Broadband electrical characterization of multiwalled carbon nanotubes and contacts. *Nano Letters*, 1086–1090, 2007.

12. S. C. Jun, J. H. Choi, S. N. Cha, C. W. Baik, S. Lee, H. J. Kim, J. Hone, and J. M. Kim. Radio-frequency transmission characteristics of a multiwalled carbon nanotube. *Nanotechnology*, 18-255701, 2007.

13. M. Zhang, X. Huo, P. C. H. Chan, Q. Liang, and Z. K. Tang. Radio-frequency characterization for the single-walled carbon nanotubes. *Applied Physics Letters*, 88, 163109, 2006.

14. J. J. Plombon, K. P. O'Brien, F. Gstrein, V. M. Dubin, and Y. Jiao. High-frequency electrical properties of individual and bundled carbon nanotubes. *Applied Physics Letters*, 90, 063106, 2007.

15. Y. Xu, A. Srivastava, and J. M. Marulanda. Emerging carbon nanotube electronic circuits, modeling and performance. *Proceedings of the 51st Midwest Symposium on Circuits and Systems*, Knoxville, TN, August 10–13, 2008, pp. 566–569.

16. M. Zhang, X. Huo, P. C. H. Chan, Q. Liang, and Z. K. Tang. Radio-frequency transmission properties of carbon nanotubes in a field effect transistor configuration. *IEEE Electron Device Letters*, 27(8), 668–670, 2006.

17. S. Bhattacharya, S. Das, and D. Das. Stability analysis in CNT and GNR interconnects. *Proceedings of the 2nd International Conference on Advanced Communication Systems and Design Techniques*, Haldia, India, September 20–30, 2012, pp. 64–66.

18. D. Fathi and B. Forouzandeh. A novel approach for stability analysis in carbon nanotube interconnects, *IEEE Electron Device Letters*, 30(5), 475–477, 2009.

19. S. Bhattacharya, S. Das, and D. Das. Analysis of stability in carbon nanotube and graphene nanoribbon interconnects. *International Journal of Soft Computing and Engineering*, 2(6), 325–329, 2013.

20. S. H. Nasiri, M. K. Moravvej-Farshi, and R. Faez. Stability analysis in graphene nanoribbon interconnects. *IEEE Electron Device Letters*, 31(12), 1458–1460, 2010.

21. S. Bhattacharya, S. Das, and D. Das. Analysis of stability in carbon nanotube interconnects. *Proceedings of the National Conference on Recent Advances in Applied Mathematics*, Gurgaon, Haryana, India, February 18, 2012, pp. 34–39.

Chapter 6

1. M. D'Amore, M. S. Sarto, and A. Tamburrano. Fast transient analysis of next-generation interconnects based on carbon nanotubes. *IEEE Transactions on Electromagnetic Compatibility*, 52(2), 496–503, 2010.
2. S.-N. Pu, W.-Y. Yin, J.-F. Mao, and Q. H. Liu. Crosstalk prediction of single- and double-walled carbon-nanotube (SWCNT/DWCNT) bundle interconnects. *IEEE Transactions on Electron Devices*, 56(4), 560–568, 2009.
3. A. G. Chiariello, A. Maffucci, and G. Miano. Signal integrity analysis of carbon nanotube on-chip interconnects. *Proceedings of the IEEE Workshop on Signal Propagation on Interconnects*, Strasbourg, France, May 12–15, 2009.
4. M. D'Amore, M. S. Sarto, and A. Tamburrano. Transient analysis of crosstalk coupling between high-speed carbon nanotube interconnects. *Proceedings of the IEEE International Symposium on Electromagnetic Compatibility*, Detroit, MI, August 18–22, 2008.
5. D. Rossi, J. M. Cazeaux, C. Metra, and F. Lombardi. Modeling crosstalk effects in CNT bus architectures. *IEEE Transactions on Nanotechnology*, 6(2), 133–145, 2007.
6. D. Das and H. Rahaman. Crosstalk analysis in carbon nanotube interconnects and its impact on gate oxide reliability. *The 2nd Asia Symposium on Quality Electronic Design*, Penang, Malaysia, August 3–4, 2010.
7. D. Das and H. Rahaman. Analysis of crosstalk in single- and multiwall carbon nanotube interconnects and its impact on gate oxide reliability. *IEEE Transactions on Nanotechnology*, 10(6), 1362–1370, 2011.
8. D. Das and H. Rahaman. Crosstalk overshoot/undershoot analysis and its impact on gate oxide reliability in multi-wall carbon nanotube interconnects. *Journal of Computational Electronics*, Springer, 10(4), 360–372, 2011.
9. D. Das and H. Rahaman. Crosstalk and gate oxide reliability analysis in graphene nanoribbon interconnects. *The 2nd International Symposium on Electronic System Design*, Kochi, Kerala, India, December 19–21, 2011.
10. H. J. Li, W. G. Lu, J. J. Li, X. D. Bai, and C. Z. Gu. Multichannel ballistic transport in multiwall carbon nanotubes. *Physical Review Letters*, 95, 086601–4, 2005.
11. L. Wilson. International Technology Roadmap for Semiconductors (ITRS) reports, http://www.itrs.net/reports.html, 2006.
12. Yu (Kevin) Cao. Predictive Technology Model, http://ptm.asu.edu/, 2008.
13. W. R. Hunter. The statistical dependence of oxide failure rates on vdd and, tox variations, with applications to process design, circuit design, and end use. *Proceedings of the IEEE 37th Annual International Reliability Physics Symposium*, San Diego, CA, March 23–25, 1999.
14. N. S. Nagaraj, W. R. Hunter, P. Balsara, and C. Cantrell. The Impact of inductance on transients affecting gate oxide reliability. *Proceedings of the 18th International Conference on VLSI Design* held jointly with *4th International Conference on Embedded Systems Design*, 2005.
15. J. F. Chen, J. Tao, P. Fang, and C. Hu. Performance and reliability comparison between asymmetric and symmetric LDD devices and logic gates. *IEEE Journal of Solid-State Circuits*, 34(3), 367–371, 1999.
16. L. Jia and W.-Y. Yin. Temperature effects on crosstalk in carbon nanotube interconnects. *Proceedings of the Asia-Pacific Microwave Conference*, Macau, December 16–20, 2008.

17. P. Sun and R. Luo. Analytical modeling for crosstalk noise induced by process variations among cnt-based interconnects. *Proceedings of the IEEE International Symposium on Electromagnetic Compatibility*, Austin, TX, August 17–21, 2009.
18. W. C. Chen, W.-Y. Yin, L. Jia, and Q. H. Liu. Electrothermal characterization of single-walled carbon nanotube (SWCNT) interconnect arrays. *IEEE Transactions on Nanotechnology*, 8(6), 718–728, 2009.
19. S. Pasricha, F. J. Kurdahi, and N. Dutt. Evaluating carbon nanotube global interconnects for chip multiprocessor applications. *IEEE Transactions on Very Large Scale Integration (VLSI) Systems*, 18(9), 1376–1380, 2010.
20. D. Das and H. Rahaman. Delay uncertainty in single- and multi-wall carbon nanotube interconnects. *Proceedings of the 16th International Symposium on VLSI Design and Test*, Lecture Notes in Computer Science 7373, Springer, Shibpur, India 289–299, July 1–4, 2012.

Chapter 7

1. L. Wilson. International Technology Roadmap for Semiconductors (ITRS) reports, http://www.itrs.net/reports.html, 2006.
2. W. Steinhögl, G. Schindler, G. Steinlesberger, and M. Engelhardt. Size-dependent resistivity of metallic wires in the mesoscopic range. *Physical Review B*, 66, 075414, 2002.
3. H. Li, C. Xu, N. Srivastava, and K. Banerjee. Carbon nanomaterials for next-generation interconnects and passives: physics, status, and prospects. *IEEE Transactions on Electron Devices*, 56(9), 1799–1821, 2009.
4. M. Budnik, A. Raychowdhury, and K. Roy. Power delivery for nanoscale processors with single wall carbon nanotube interconnects. *Proceedings of the 6th IEEE Conference on Nanotechnology*, Cincinnati, OH, June 17–20, 2006, pp. 433–436.
5. A. Naeemi and J. D. Meindl. Performance Modeling for Single- and Multiwall Carbon Nanotubes as Signal and Power Interconnects in Gigascale Systems, *IEEE Trans. Electron Devices*, 55(10), 2574–2582, 2008.
6. D. Das and H. Rahaman. IR drop analysis in single- and multiwall carbon nanotube power in sub-nanometer designs. *Proceedings of the 3rd Asia Symposium & Exhibits on Quality Electronic Design*, Kuala Lumpur, Malaysia, July 19–20, 2011, pp. 174–183.
7. H. H. Chen and D. D. Ling. Power supply noise analysis methodology for deep-submicron VLSI chip design. *Proceedings of the 34th Design Automation Conference*, Anaheim, CA, June 9–13, 1997.
8. K. Arabi, R. Saleh, and X. Meng. Power supply noise in SoCs: Metrics, management, and measurement. *IEEE Design and Test of Computers*, 24(3), 236–244, 2007.
9. S. Lin and N. Chang. Challenges in power-ground integrity. *Proceedings of the International Conference on Computer-Aided Design*, 2001.
10. Yu (Kevin) Cao. Predictive Technology Model, http://ptm.asu.edu/, 2008.
11. K. T. Tang and E. G. Friedman. Simultaneous switching noise in on-chip CMOS power distribution networks. *IEEE Tranactions on Very Large Scale Integration (VLSI) Systems*, 10(4), 487–493, 2002.

12. D. Das and H. Rahaman. Simultaneous switching noise and IR drop in graphene nanoribbon power distribution networks. *Proceedings of the 12th International Conference on Nanotechnology*, Birmingham, August 20–23, 2012, pp. 1–6.

13. Q. Shao, G. Liu, D. Teweldebrhan, and A. A. Balandin. High-temperature quenching of electrical resistance in graphene interconnects. *Applied Physics Letters*, 92, 202108-1–202108-3, 2008.

14. S. Ghosh, W. Bao, D. L. Nika, S. Subrina, E. P. Pokatilov, C. N. Lau, and A. A. Balandin. Dimensional crossover of thermal transport in few-layer graphene materials. *Nature Materials*, 9(7), 555–558, 2010.

15. A. Naeemi and J. D. Meindl. Compact physics-based circuit models for graphene nanoribbon interconnects. *IEEE Transactions on Electron Devices*, 56(9), 1822–1833, 2009.

16. D. Das and H. Rahaman. Modeling of IR-drop induced delay fault in CNT and GNR power distribution networks. *Proceedings of the 5th International Conference on Computers and Devices for Communication*, Kolkata, India, December 17–19, 2012.

Index

Note: Locators followed by *"f"* and *"t"* denote figures and tables in the text